Bioethics and Racism

Bioethics and Racism

Practices, Conflicts, Negotiations and Struggles

Edited by
Carlo Botrugno, Marcia Mocellin Raymundo and Lucia Re

DE GRUYTER

ISBN 978-3-11-144775-9
e-ISBN (PDF) 978-3-11-076512-0
e-ISBN (EPUB) 978-3-11-076516-8

Library of Congress Control Number: 2022946793

Bibliographic information published by the Deutsche Nationalbibliothek
The Deutsche Nationalbibliothek lists this publication in the Deutsche Nationalbibliografie;
detailed bibliographic data are available on the internet at http://dnb.dnb.de.

www.degruyter.com

Table of Contents

Carlo Botrugno, Marcia Mocellin Raymundo, Lucia Re

Introduction

Bioethics emerged in the middle of the 20th century, mostly as a reaction to racist crimes perpetrated under the auspices of healthcare, such as the atrocities of Nazi medical experimentations or the outrageous Tuskegee Syphilis study conducted in the US from 1932 to 1972. Since then, the field of bioethics has grown notably, expanded its domain and strengthened its toolbox, thus contributing to the enhancement of patients' rights, dignity, and autonomy and their awareness of the medical treatments they undergo. Indeed, bioethics today is a flourishing discipline and currently serves as a "bridge to the future" as in the pioneering view of Van Rensselaer Potter[1], who understood the need for a new "wisdom" – and thus for a new discipline – located at the intersection between biological knowledge and human values.

The development of bioethics today is inextricably linked to the magnitude and the relevance of current technological advancements in medicine and related fields. These include recent discoveries in the field of genetics and its subfields as well as new technology-driven medical treatments, which have amplified conflicts of values and interests and made it increasingly complex to find an adequate balance among them. The need to address these challenges strengthens the value of the knowledge and the tools offered by bioethics to both healthcare professionals and patients.

However, despite the fact that discrimination and racism in healthcare were the foundation of bioethics, today they are considered to be "minor" research interests. In less than a century, the emphasis of bioethics scholars on the "cutting-edge" and the "technology-driven" has slowly allowed the relevance of cultural, social, and economic factors in health and healthcare to fade into the background. This was a primary concern for Giovanni Berlinguer, an Italian physician who coined the concept of "everyday bioethics"[2] in opposition to "frontier bioethics"; the latter is predominantly concerned with technological advancements and the latest discoveries, while the former embraces issues that involve the majority of people in contemporary societies, such as healthcare access, equity in health resources allocation, health inequalities and discrimination, public health, and primary care services.

---

1 Potter, Van Rensselaer. (1971): *"Bioethics: Bridge to the future.* Englewood Cliffs: N. J. Prentice-Hall.
2 Berlinguer, Giovanni. (2000): *Bioetica quotidiana.* Firenze: Giunti; (2003): *Everyday Bioethics: Reflections on Bioethical Choices.* New York: Baywood Publisher.

https://doi.org/10.1515/9783110765120-001

In advocating for an everyday bioethics, Berlinguer did not mean to deny or underestimate the relevance of technological advancements and new discoveries in medicine, nor did he have in mind a new or alternative theory of bioethics. Rather, he advocated for a perspective that would be able to revitalize the importance of elements such as discrimination, racism, and (in)equity, which have always belonged to the domain of bioethics but have been overshadowed by the prevalence of more individual-based theories, approaches, and research topics. In his "Bioética Cotidiana"[3], Berlinguer started with a personal anecdote about a trip to attend a conference in Brazil. He reported that, while he was on the plane, he realized that many old white men were travelling from Europe to Latin America solely for the purpose of sex tourism. In his mind, this episode triggered – furthermore – the urgency of strengthening the connection of bioethics with global social justice, which is the core element of his everyday bioethics.

Although it may seem that social justice is unrelated to health and healthcare, inequalities and discrimination exert a significant influence on both, especially when referring to the most vulnerable population groups. Berlinguer did argue that unveiling and fighting this hidden racism is a major goal of a bioethics that contributes to reversing the direction of our racist societies, promoting equality and respect, and thus defending everyone's right to exist.

To strengthen the connection with social justice, bioethics must aim at "weighing the difference"[4], i.e. considering the impact of cultural, social, and economic factors – including violence, racism, and discrimination – on health and healthcare access. This requires dedicating more effort to researching the mechanisms through which hegemonic power relations condemn certain population groups to a condition of subjugation, suffering, and oppression.

On a closer look, this proposal fits the idea of a "critical bioethics" or a "bioethics of the global south" that has emerged over recent years, especially in Latin America and the Caribbean[5]. The conceptualization of critical bioethics is based on an acknowledgement of deep inequalities in health conditions and healthcare access on the part of local populations. Likewise, this proposal also highlights the need to re-elaborate the ethical responsibilities of the interna-

---

3 Berlinguer, Giovanni. (2004): *Bioética Cotidiana*. Brasília: Editora UnB.
4 Botrugno, Carlo. (2018): "Everyday Bioethics, Migrations, and Healthcare: Weighing the difference". *L'Altro Diritto* 1. pp. 78–101.
5 García Alarcón, Rodrigo Hernán. (2012): "A bioética na perspectiva latino-americana, sua relação com os direitos humanos e a formação da consciência social de futuros profissionais". *Revista Latinoamericana de Bioética* 12. N. 2, pp. 44–51. Lorenzo, Cláudio; Garrafa, Volnei (2011): "Ensayos clínicos, Estado y sociedad: ¿dónde termina la ciencia y empieza el negocio?". *Salud Colectiva* 7. N. 2, pp. 166–170.

tional community in situations in which the structural mechanisms from which healthcare injustice arises are left unaltered[6].

This volume is based on these theoretical premises and aims to explore the links between racism and bioethics, paying particular attention to the epistemic, social, legal and political dimensions of these links. It is rooted in a discussion and dissemination initiative organized by the RUEBES (Research Unit on Everyday Bioethics and Ethics of Science) in 2021. The RUEBES was created in 2016 within the L'Altro Diritto Research Centre at the University of Florence[7] to foster "everyday bioethics". Since then, the RUEBES has been working in several directions, but always with the purpose of intensifying the debate and increasing people's awareness of neglected rights, abuses, discrimination and other forms of oppression and prevarication in healthcare and health-related fields. Today, the RUEBES includes young and motivated scholars from several countries and from a variety of disciplinary backgrounds – law, sociology, anthropology, health sciences, bioethics, medical ethics, philosophy, etc. The enthusiasm and efforts of this group led to the organization of the "Bioethics in Action Seminar series" (BiASs), which reached its third edition in 2022. The BiASs has brought together many international scholars to discuss, share, confront, and raise awareness on topics such as commodification in healthcare, disability, health literacy, healthcare inequalities, mental health, human rights and bioethics, racism and discrimination, environmental health justice, capitalism and care, genomics and genetics, conflicts of interest in biomedical research, etc. In particular, the second edition was fully dedicated to the topic of "racism and bioethics", a research domain that has been largely overlooked – as Berlinguer already remarked many years ago. Throughout the BiASs, we have developed interdisciplinary and international opportunities for dialogue concerning "practices, policies, negotiations and struggles" pertaining to bioethics and racism.

Many of the scholars – young and senior – who participated in BiASs II agreed to contribute to this collective volume. In a (not so) strange parallel with Berlinguer's inspiring journey to Brazil, the RUEBES activity and this book in particular have taken advantage of many fruitful connections with Brazilian scholars and research institutions. Similarly, some of the chapters included in this work connect in several ways to the bioethical issues that are currently at stake on the African continent. Others cover topics whose interest is unrelated

---

6 Buxó i Rey, Maria. (2004): "Bioética Intercultural para la Salud Global". *Revista de Bioética y Derecho* 1. Pp. 12–15.
7 See the RUEBES website at http://www.adir.unifi.it/ruebes/.

to a peculiar geographical area. All of them, however, have a global mission and result from a successful crossing of disciplinary boundaries, perfectly matching the RUEBES mission statement.

The book opens with the essay by Lucia Re, who adopts Michel Wieviorka's and Michel Foucault's perspectives on racism, considering it to be a feature of modern societies and trying to show how it persists in contemporary ones with a structural character. Some of the main manifestations of structural racism in the United States, Brazil and Europe are examined in detail. Moreover, following Achille Mbembe's arguments, the essay shows how racism preserved the domination of certain social groups in the transition from the colonial to the post-colonial era and how it is necessary today to establish a regime of inequality both globally and within (neo)liberal-democracies. In light of her analysis, the author finally questions the effectiveness of anti-racist discourse and affirmative actions in contemporary (neo)liberal democracies and proposes that they are connected to a strong revival of 20th century "(inter)national constitutionalism", which focuses on substantial equality and the equal recognition of differences.

In the second chapter, Carlo Botrugno emphasizes that racism is deeply rooted at epistemic level, i.e. in the production of knowledge relevant to bioethics and related fields, as it is veiled in policies, inscribed in social and professional practices, and infused in people's views and perspectives, often in an unconscious and unintended way. Based on the work of Boaventura de Sousa Santos, Botrugno emphasizes the "racial epistemicide" of bioethics by focusing on some concrete manifestations of it. From this perspective, the emergence of "whiteness" itself as a category through which to identify and fight racism in bioethics must be viewed as informed by a (racist) attempt to categorize people on the basis of their skin color. Although the biological notion of race has already been dismantled and relegated to a social construct or a racist metaphor, the relationship between race and genetics continues to exert its "discreet charm" on biomedical research, and consequently also on the practice of medicine and clinical ethics. Such an epistemic critique expands to problematize the role played by medical educators as the "gatekeepers" of bioethics, particularly in the context of bioethics courses, as well as by clinical and research ethical committees. In light of this, Botrugno embraces the proposal of Katya Gibel Mevorach (also in this book) to construct new epistemic notions and categories through which to dismantle the hegemonic relationships and structures of power that made possible the epistemicide of bioethics and thereby the perpetration of racism, oppression, and social injustice.

In the third chapter, Robin Pierce investigates the role of technology in the creation and perpetuation of health inequities in the provision of healthcare, which must be viewed as an "urgent need" in the context of an increasing digitalized society, particularly when considering the current pressure towards the digitalization of healthcare globally. After acknowledging that the benefits of technological innovation have resulted in improved outcomes of multiple sorts, Pierce notes that the degree to which technology confers benefits is not experienced equally across groups and that health disparities along racial lines have been documented for decades. These inequities are today "inexcusable" and are increasingly viewed as a public health concern. In light of this, Pierce recognizes the potential usefulness of regulation theory, particularly with respect to technology, in developing effective strategies to interrupt technological pathways to health inequities and address its root causes.

In the fourth chapter, Celia Mariana Barboza de Sousa, Fernanda Sales Luiz Vianna, and Francisco Veríssimo Veronese trace a connection between the processes of marginalization and segregation as experienced by the Afro-Brazilian population during colonization and the health needs of the black population by focusing on chronic kidney disease (CKD), a condition that can lead to a decrease in kidney function and the consequent need for dialysis. As they emphasize, CKD is more concerning among people with an African heritage due to its higher prevalence in black populations. In light of this, they advocate for the importance of carrying out studies focusing on population aspects of African and admixture groups, given their potential to improve strategies for CDK prevention and therefore public health policies aimed at protecting and benefitting the black population.

In the fifth chapter, Nchangwi Syntia Munung problematizes the use of population descriptors such as race, ethnicity, geographical location, and ancestry as analytical variables in biomedical research. As complex social constructs, these descriptors may include stigmatizing undertones that have been "naturalized" over time. Therefore, their current use – Munung argues – should be justified by scientific reasons and thus assessed in light of their impact, i.e. their social acceptability. This process would require science outreach activities to consult with different stakeholders regarding what each population descriptor means to them, particularly in cases in which a descriptor may have pejorative undertones because of its historical underpinnings or when the focus of the study is on a disease that is stigmatized, thus indicating that the use of a particular descriptor may further perpetuate stigma and/or discrimination.

In the sixth chapter, Katya Gibel Mevorach begins with the question "what difference does difference make?" to focus on the need to intervene against and discredit the notion of "race" in science education and research. As she illustrates, the investigation of the history, context, and consequences of racism reveals the necessity of vigilance in teaching and the need to uproot false ideas that can otherwise serve as the premise and point of departure for racism-linked prescriptive questions and interpretations. For Mevorach, this entails candidly confronting and dismantling racism and highlighting its effects. This means that, in the area of medical research, diagnosis and treatment, categorization of human beings that relies on a racial taxonomy, racial categories and racial derivatives should be relegated to the historiography of scientific racism and viewed as "bad science". From this perspective, human variation based on subjective and learned differences are rooted in political, economic, and legal reasons and histories, although they have biological consequences that impact the health and well-being of humans. In light of this, scientists should not deemphasize the weight of the legacies of discriminatory racist legislation and policies in their research. This should cause them to commit to drafting new data schemes – or "curating a new syllabus", in the words of Mevorach – in research projects about humans to unlearn the biases of the societies in which they live and reject the biases of the academic environment in which they work.

In the seventh chapter, Cláudio Lorenzo and Marcia Mocellin Raymundo discuss some examples of structural and institutional racism in the field of healthcare, seeking to establish relationships with bioethics and human rights by examining the particular situation of Brazil. As they remind us, today, many healthcare professionals encounter great difficulty understanding and respecting cultures other than their own. Due to its ability to support coexistence, enhance citizenship, and defend human rights, bioethics can provide useful tools for conflict resolution and encourage respect for different worldviews in intercultural societies. In such a context, the authors exhort bioethics and those who work in this field to address the issue of racism not only in a theoretical context but above all in everyday practice.

In the eighth chapter, Josimário João da Silva and Ronaldo Piber discuss the term "misthanasia" (*mis* = unhappy; *thanatos* = death), which means "unfortunate death", to account for the current situation of Brazil, where a large part of the population is condemned to death by conditions of poverty as well as by the spread of violence, lack of infrastructure, and absence of the minimum conditions for decent living. From the perspective of these authors, therefore, dying does not merely coincide with physical death; rather it starts from the moment in which people are denied their fundamental rights and, above all, access to the minimum conditions of citizenship, which must include the right to health. In such a context, the authors remark that the Federal Constitution of Brazil identifies the right to health as a fundamental right and imposes a duty to protect it on the state itself. However, they complain about the inertia of public institutions with respect to increasingly common phenomena of racism and discrimination throughout the country.

In the ninth chapter, Inês Faria, Laura Brito, and Karla Costa discuss episodes and data drawn from studies pertaining to two countries with a colonial relationship, i. e. Mozambique and Portugal, with the ultimate purpose of deconstructing the history of discrimination and prejudice in the field of reproductive care. By exploring regimes of invisibility and forms of structural violence concerning black women, they argue for the need to overcome rationales based on stereotypes concerning race, gender, and class that are prevalent in reproductive care settings. The persistence of these rationales highlights the lingering effects of colonial history and the practices of subalternization that lead to racism and discrimination in healthcare (among others). In such a context, the authors emphasize the need to promote horizontal conversations and engage in cross-sectorial and multidisciplinary dialogues through which everybody, including citizens, researchers, practitioners, and patients, can contribute to provoke such a change.

In the tenth chapter, Judith van de Kamp focuses on the interactions and power dynamics between medical staff and students of international medical electives, on the one hand, and students from a rural hospital in Northwest Cameroon, on the other. Her ultimate purpose is to contribute to addressing the knowledge gap related to the challenges resulting from international students' misunderstandings of situations at the medical workplace, which lead to all kinds of "processes of othering" as coping mechanisms, often with the effect of – intentionally or unintentionally – creating distance between "us" and "them". From the perspective of van de Kamp, this hinders the development of strong relationships with the health workers involved, both at the individual and the institutional level. In light of this, she encourages education institutes in high-income countries to fulfill their responsibility to prevent harmful interactions and improve their international internship support programs.

Finally, in the eleventh chapter, Carlo Botrugno discusses the potential contribution of bioethics to the task of addressing inequalities and discrimination in healthcare, focusing on undocumented migrants and asylum-seekers in destination societies. As he argues, at the heart of these phenomena lie xenophobic and racist attitudes, practices and policies. After outlining the main threats to the health of migrants throughout their illegal journeys as well as the main obstacles that usually prevent them from accessing adequate healthcare services in their destination countries, the author draws attention to the impact of COVID-19 pandemic on migrant populations, particularly in light of the growing push toward the digitalization of healthcare and its potential to create further barriers to service accessibility. He then discusses – and rejects – the association between migrations and the spread of infectious diseases, which is often used to feed political tensions regarding the management of international migration flows as well as to implement racist policies that prevent migrants from having full access to healthcare in destination countries.

The main aim of the essays collected here is thus to shed light on some of the practices, conflicts, negotiations and struggles related to bioethics and racism in the hope that there will be a return to a strong focus on this crucial field of research. If bioethics is to combat racism, discrimination, and inequalities, more efforts are warranted with respect to researching the mechanisms through which hegemonic power relations condemn some population groups to a condition of subjugation, suffering, and oppression. However, by unpacking certain notions that have been taken for granted and dismantling rhetorics that are veiled in discourses and rationales pertaining to race and racism in the context of health and healthcare, we highlight possible ways in which bioethics can operate across disciplinary boundaries and strengthen its connection with equity and social justice, which also entails striving for a "bioethics in action".

Bhuiyan Md Sohrab Hossain
Peddler

Sufferings needed? Sufferings!
Various types of sufferings
Sufferings needed? Sufferings!
Red distress, blue agony, bright yellow coloured pains
Pallid pain of green grass pressed by stone
Dark ache of light
I've multi-coloured anguish
Sufferings needed? Sufferings!
Sufferings of the house, of the alien, of birds and leaves
Pain of chin
Pain of eyes, of bosom, of nails
I've sufferings of a man who gets lost gradually
Sufferings needed? Sufferings!
Pain of love, of hatred, of rivers, and pain of the damsel
Severe pain of neglect and humiliation
Sufferings of loving an unknown woman
Of gatherings of corrupt leader
I've sufferings of devastation by hydrogen
Sufferings needed? Sufferings!
Sufferings of the daytime, pains of night
Pains of the path of the feet
Immense pathetic sufferings, sufferings of the hawkers
Sufferings needed? Sufferings!
Nobody ever gives you except me
Real artistic sufferings
The life-long sufferings of someone else
Who has ever destroyed everything like me?
Who can ever offer healthy agony except me?

https://doi.org/10.1515/9783110765120-002

Lucia Re
Reproducing Hierarchies. Racism from Colonialism to Neoliberalism

Abstract: In this essay, I adopt the perspectives of Michel Wieviorka and Michel Foucault, which consider racism to be a feature of modern societies, and I try to show how it persists in contemporary ones, having a structural character. I then examine in more detail some of the main manifestations of structural racism in the United States, Brazil and Europe. Moreover, following Achille Mbembe's arguments, I discuss how racism maintained the domination of certain social groups in the transition from the colonial to the post-colonial era and how it is necessary today to build up a regime of inequality both globally and within (neo)liberal-democracies. Finally, I argue that, within this framework, anti-racist discourse, especially that promoted by supranational and international institutions, often ends up having a limited impact. On the other hand, affirmative actions promoted by discriminated groups are frequently insufficient. To be effective, both anti-racist and affirmative policies must then be linked to a strong revival of 20th century "(inter)national constitutionalism", which focuses on substantial equality and the equal recognition of differences.

1 Introduction

The term "racism" has been used to explain different social phenomena. I adopt here Michel Wieviorka's (1998) perspective, which considers it to be a feature of modern societies, and I try to show how it persists in contemporary ones. From this point of view, racism is not only the expression of hostility toward populations that are labelled as "different" but rather a tool that allows to rebuild hierarchies in societies based on individualism, in which status is no longer taken for granted. Racism is used to delimit "groups", "communities", "nations", "peoples", etc. and to order them according to a superior/inferior logic. It is thus, as Michel Foucault (2011) has noted, a device that is at work in a multiplicity of narratives, policies and practices and that consists in an intertwining of power and knowledge. It is not a mere belief or ideology. It has a structural character. Sometimes it is explicit, while at other times it is denied or overlooked but continues to operate, producing multiple forms of discrimination and abuse.

Lucia Re, University of Florence

https://doi.org/10.1515/9783110765120-003

In this essay, I consider some of the main manifestations of structural racism in the United States and Europe. I also refer to South America, especially to Brazil, where the structural dimension of racism is particularly manifest. Following Achille Mbembe's (2016a; 2016b; 2020) arguments, I then discuss how racism not only has a persistent and, indeed, structural character but was used to maintain the domination of certain social groups in the transition from the colonial to the post-colonial era and is necessary today to build up a regime of inequality both globally and within (neo)liberal-democracies. Indeed, the neoliberal order – based on the myth of individual success and market competition – needs to preserve and strengthen social hierarchies.

I use the term "neoliberalism" according to the broad definition given by Wendy Brown (2015): neoliberalism is the project that promoted the rise of a "rationality extending a specific formulation of economic values, practices and metrics to every dimension of human life". It involves the "'economization' of heretofore noneconomic spheres and practices (...)". Such a political project – which has been carried out globally from the second half of the 1970s onward – leads to a progressive expansion of inequality and results in the precarization and vulnerability of ever larger segments of the world population. It erodes the protections that 20th century welfare states used to guarantee to their citizens, fueling in the population that has hitherto benefited, whether consciously or unconsciously, from "white privilege" the fear of sharing the fate of people who for centuries have been regarded as belonging to "inferior races" and who have been exploited and/or marginalized.

The fear of losing one's "security" fuels "dirty racism" – a form of popular, explicit racism – that reinforces structural racism. The need to mark the boundary between the included and the excluded grows stronger in neoliberal societies. It activates the "scapegoat" dynamic (Girard 1982) and leads to the adoption of policies designed to reassure the "white" population[1]. These policies

---

1 The terms "race", "white", "black", etc., employed in this essay must be considered as constantly written in quotation marks. They refer to hetero- or, more rarely, self-ascriptions that have historically been attributed to certain groups on the basis of somatic traits, national origin, social affiliation, etc. In Europe, the use of these terms is mostly avoided, and the category of "race" is evoked, as I will also do, only in negative terms to detect discrimination against certain social groups. In the United States – and to some extent in Brazil – the reference to the term "race" and/or the "racial" categories of "black", "white", "negro", "pardo", etc. is more common and is also found in official documents, both because the history of these countries is different from the history of Europe, where racial categories were used by totalitarian regimes to justify the extermination of Jews and Roma, and because in these countries there has long been, as I will discuss, a re-appropriation of "racial labels" by discriminated groups. These labels are thus directed against "color-blind" policies that, by ignoring it, perpetuate discrimination.

affect the life of members of discriminated groups. Within this framework, anti-racist discourse, especially that promoted by supranational and international institutions, predominantly focuses on more explicit forms of racism, neglecting the structural mechanisms that perpetuate the inferiorization of the largest part of humanity. It thus ends up having a limited impact.

On the other hand, discriminated groups fight a fundamental battle for the recognition of structural racism and demand policies to rebalance and repair the injustice and violence suffered in the past and still perpetrated against them today. They often reappropriate their racialized identity, calling for affirmative actions. As I argue, policies that counter color blindness and promote the protection of identity rights play a key role in combating both voluntary and structural racism. To be effective within the framework of neoliberalism, however, they must be linked to a robust revival of the project of 20th century "(inter)national constitutionalism" (Mazzarese 2018). That political "project of equality" (Ferrajoli 2018), which arose after the Shoah and at the beginning of decolonization movements, focusing on substantial equality and the equal recognition of differences, can counter the political "project of inequality" promoted by contemporary neoliberalism and revive a political and legal struggle against structural racism and the practices of dehumanization that are associated with it.

Universalism has historically been used in an "ethnocentric" way to promote the superiority of "Western culture" (Zolo 2000; Re 2012). And yet, the notion of common membership in the human species is an increasingly necessary idea, not only to counter racism but also for the survival of human beings on Earth in the face of ecologic crisis (cf. Volpato 2014). In addition to this idea, there should be an awareness of the vulnerability of other living species and the ecosystem in which we live. What we need, then, is not a top-down universalism that is used *ex parte principis* to reinforce domination but rather a bottom-up universalism, conceived of as a "universality of multiplicity" (Pitch 2004), which emerges from a necessarily plural experience and remains open to welcoming and accommodating different and changing identities.

2 Structural racism

Racism certainly has more than one root. George L. Mosse (1978) has shown this very clearly in his famous history of racism in Europe, investigating its theoretical and religious sources and highlighting its rise in science and art. As Mosse himself noted, however, modern racism is closely related to nationalism. Processes of othering are found in various forms of nationalism, as they serve to delimit the boundaries of a political community on the basis of belonging to a "people" imagined as "ethnos". This idea of the "people" fuels the nationalist rhetoric of many European countries (cf., e. g., Mosse 1978; Noiriel 2007; Rigione 2020)[2], while it is less manifest in the nationalist narrative of the United States. However, it was precisely in the United States that the system of the racial segregation of blacks lasted for a long time, thus creating a "racial divide" that can still be detected today. Racism has been present in the United States since its origin. It affects not only the people who have been enslaved but also native peoples. The image of the North American nation is, indeed, nourished by the colonial myth of the frontier, a myth that is heavily imbued with racism.

As Bartolomé Clavero (2002, pp. 537–538) noted, the constitutional history of the American continent was inaugurated by the United States with the exclusion of native peoples from the constitutional compact. Native tribes were viewed in the Federal Constitution of 1787 as a "third kind", neither a foreign nation nor an integral part of the state. They were considered to be "domestic dependent nations" placed in a position of inferiority to the population of European origin. This colonial legacy is intertwined with processes of racialization and inferiorization.

The colonial conditions of reservation and "tutelage" (Clavero 2002, p. 540) for native peoples has persisted over time. Native peoples have been confined to territories that have been granted very limited autonomy. Subsequent legal and political developments have thus led to the constitutionalization of "an essentially colonial position"[3] (Clavero 2002, p. 541). This strong colonial legacy is interwoven with the legacy of slavery and segregation, which are also of colonial origin. Angela Davis masterfully illustrated it in "Women, Race & Class" (1981), showing the intertwining between sexism and racism and enlightening the ambiguous positions towards black people taken by some white members of antislavery and early feminist movements.

As immigration from Europe increased during the 19th and 20th centuries, immigrants were also categorized on racial grounds. Some of them – e. g., Italians coming from the south of the peninsula – were considered, along with the Portuguese, Spanish, French, Greek, North African and Middle Eastern immigrants, to be of a "Mediterranean race" and labelled as inferior to the "WASP"[4] population (Palidda 2008, p. 10). They often shared the same fate as freed African-Americans, suffering from very harsh living and working conditions (cf. Davis 1981), and in states like Alabama, Georgia and Louisiana, they experienced different

---

2 We will return on their specific forms of racist nationalisms in § 3.
3 Translation from Italian by L. Re. In the following, all quotations from texts published in languages other than English should be understood as translated by Re.
4 White-Anglo-Saxon-Protestant.

forms of legal segregation. As Palidda (2008, p. 10) recalled, "Upon arrival at Ellis Island (New York), Italians were separated and registered in two different records: Mediterranean on one side and Nordic on the other"[5]. The Dilligham Commission of the U.S. Senate ratified these racial categories in 1911, arguing in its report for the relevance of "black genetic traits" in the Neapolitan and Sicilian people, not only with respect to explaining their physical appearance but also their personalities and inclinations[6]. Palidda, mentioning Franzina and Stella (2002), also recalls how, in the southern states of the union, Italian immigrants were called "black daggers" to indicate that they were similar to blacks and that they were criminals. They were subjected to both segregationist policies and attacks by white supremacists. This racialization was also present in other areas of the world to which Italians emigrated, particularly in France (Palidda 2008, p. 11).

In the United States, immigrants have continued to be "filtered out" over time according to racial criteria (cf. King 2002) and have been permitted a differentiated access to citizenship (Jacobson 1998). Even today, beyond the rhetoric of the "nation of immigrants" (Kennedy 1964), the hierarchies of origin can be seen in their full harshness in the various forms of social marginalization and exposure to violence that affect people who belong to racial minorities. In particular, native, black, and Latino communities are not only discriminated against because of their skin color. They are in a condition of structural subordination that is directly derived from different colonial legacies: the colonization of North America, the African slave trade, the predatory conquest of South America.

The myth of a cohesive community that identifies itself in terms of its opposition to "others" is also found in the history of many Central and South American states, whose political, legal and social systems were based on distinguishing between citizens of European origin and natives, even if, in most cases, indigenous people were formally considered to be members of the same nation by the constitutions adopted from the 19th century onward (Clavero 2002). In Brazil, black scholars have challenged the myth of *miscigenação* fueled by dominant 20th century sociology, according to which the Brazilian population is so mixed that it is impossible to identify distinct racial groups, and Brazil is therefore a state in which "color" is not an issue. Instead, they denounced the structural racism that still shapes Brazilian society, where blacks are the majority of the pop-

---

5 To read the labels on the certificates, cf. https://www.statueofliberty.org/ellis-island/family-history-center/, visited on 7 July 2022.
6 Text quoted in Italian in Palidda 2008, p. 10; original source at https://curiosity.lib.harvard.edu/immigration-to-the-united-states-1789-1930/catalog/39-990014299020203941, visited on 7 July 2022.

ulation but are underrepresented in education, politics, the media and, more generally, the ruling class, while they are overrepresented in the poorest and most marginalized sections of the population. Djamila Ribeiro (2019, p. 24), recalling these analyses, spoke of an "epistemicide", that not only happened historically but continues today to annihilate the cognitive capacity and self-esteem of the black population, both through a discriminatory school system and through the invisibilization of blacks as knowledge-producing subjects. This process leads to a "cultural whitening" that alienates blacks from culture and education, blocking their access to the "social elevator". Such "symbolic violence" (Bourdieu 1980) has been accompanied by policies of outright racial selection, such as massive sterilization of black women.

In 1992, a Parliamentary Inquiry Commission was established in the Brazilian Congress to investigate the use of sterilization as a method of contraception, since such an invasive technique was the most widely used in the country. The phenomenon was extremely broad. Suffice it to say that at the time of the inquiry, as many as 45 % of Brazilian women of fertile age were reported to have been sterilized. This figure was much lower in wealthier states with a higher percentage of white population. According to the Commission – which endorsed the allegations made by Brazilian anti-racist organizations – not only did black women have less access to other contraceptive methods, they were also subjected more often to involuntary sterilization. This has historically been partly due to opaque strategies used for black population control – implemented as a result of the influence of the United States, which invested in population control policies in the so-called Third World during the 1970s and 1980s – and partly due to the more precarious health conditions faced by black women, for whom sterilization was therefore recommended more often, especially after several pregnancies resulting in multiple cesarean sections (cf. Congresso Nacional 1993). Again, this policy has been widely tested on black women in the United States. Dorothy Roberts (1997), in "Killing the Black Body: Race, Reproduction and the Meaning of Liberty", asserted that the main method of birth control among the black population in the U.S. until the 1990s was sterilization through hysterectomy and later became chemical sterilization through drugs[7].

Last but not least, to highlight the structural racism that shapes contemporary societies, we must mention the selectivity of the penal and penitentiary system. This is a mechanism that has been used in different areas of the world at least since the industrial revolution (Foucault 1975). As critical criminology has long argued, "not only are the norms of criminal law selectively formed and en-

---

7 Roberts's analyses are discussed in Casalini 2007.

forced by reflecting existing unequal relations, but criminal law also exercises an active function in reproducing and producing inequality" (Baratta 1980, p. 166). Indeed, social selectivity operates at all levels: from the identification of criminal offenses within the legal system to the enforcement of prison sentences, via police stops, indictment, defense, etc. This leads to a specific social structure in which a certain population is subject to criminal and penitentiary control and, particularly, prison: such a population is, in fact, disproportionately composed of people who belong to the most marginal social classes (cf. Re 2022).

However, criminalization not only functions selectively on the basis of social class but also implements structural racial discrimination. This has been highlighted in the literature focusing on the U.S. penal and penitentiary system (Wacquant 2001; Davis 2003; Mauer 1999) but can also be detected, with some differences, in Brazil (Ribeiro 2019) and Europe (Re 2006). In the United States, the overrepresentation of African Americans in the criminal justice system, particularly among prisoners, is a historical fact. It is also recorded today, however, with respect to the Hispanic minority and has been increasing for both groups between the 1980s – when the prison population began to increase exponentially, leading to what has been described as a veritable "prison boom" (Re 2006)[8] – and 2020. At the end of 2019, there were 1,096 African American inmates per 100,000 African American residents in state and federal prisons compared to 525 Hispanic inmates per 100,000 Hispanic residents and only 214 white inmates per 100,000 white residents (U.S. Department of Justice 2020a, p. 1). In 2020, the prison population decreased by 15%, mainly due to the COVID-19 pandemic. The number of detainees belonging to racial minorities also decreased. However, even in this year, the data show significant disproportionality between the white and minority population. In particular, an estimated 2% of all black male U.S. residents and 1% of all American Indian and Alaska Native male U.S. residents were serving time in state or federal prison on December 31, 2020. Black males were 5.7 times as likely to be imprisoned in 2020 as white males; black males ages 18 to 19 were 12.5 times as likely to be imprisoned as white males of the same age (U.S. Department of Justice 2021, p. 23).

Overrepresentation is also significant with regard to women prisoners. Data show that black females (65 per 100,000) and Hispanic females (48 per 100,000) were imprisoned at higher rates than white females (38 per 100,000) in 2020. In the same year, the imprisonment rate for Native American and Alaska Native females ages 30 to 39 was more than 430 per 100,000, the highest among all fe-

---

8 Since 2009, the prison population has begun to decline, but the United States still holds the world record for the number of people incarcerated (cf. Prison Policy Initiative 2020).

males. Native American and Alaska Native females were 4.3 times as likely as white females to be in prison at the end of 2020. Females ages 18 to 19 had the highest disparity in imprisonment rates between "whites" and other "races" in 2020: native American and Alaska Native females ages 18 to 19 were 5.1 times more likely than white females of the same age to be in state or federal prisons, while the ratio was 4.1 for black females, 1.8 for Hispanic females, and 0.2 for Asians, native Hawaiians, and other Pacific Islanders (U.S. Department of Justice 2021, p. 23).

The causes of this overrepresentation of minorities in prison are multiple and depend as much on the conditions of social disadvantage that characterize these groups as on criminal policies. Certain behaviors that are exhibited most frequently by members of minorities living in poor neighborhoods, are in fact more criminalized than similar behaviors on the part of members of more affluent social classes, which are mostly composed of whites. Those who take drugs in their homes are, for example, less subject to criminal repression than those who do so on the street.

Regarding police behavior, official data show that whites, blacks and Latinos are subject to similar degrees of control in public spaces (U.S. Department of Justice 2020b, p. 1). However, the threat of the use of force is experienced more frequently by African Americans and Hispanics. Blacks and Latinos are also handcuffed more often than whites when they encounter with the police. The disproportion between the number of stops of members of minorities and those of white residents is much greater in some areas of the country where discrimination is strongly felt by black and Latino communities. In addition, research has shown that stops of members of minorities are more discretionary and lead more often to the arrest of the person being stopped (Ghandnoosh 2015).

However, more serious incidents in which black people have been killed by police officers have been the main factors affecting public opinion in recent years, sparking protest movements that have crossed U.S. borders (cf., e. g., International Commission of Inquiry on Systemic Racist Police Violence Against People of African Descent in the United States 2021). Since 2013, the Black Lives Matter movement has focused on defending African American communities from police violence and, more broadly, denouncing the racism that still pervades U.S. society[9].

The overrepresentation of minorities in prison also depends on the way in which trials and punishment are conducted. In fact, African Americans are

---

9 Cf. https://blacklivesmatter.com, visited on 6 July 2022.

kept in custody pending trial and are sentenced to prison terms more often than whites. They also receive longer prison sentences on average (National Research Council 2014, pp. 93–94). Blacks and Latinos generally have fewer resources to pay bail and defend themselves in court, fewer opportunities for social reintegration, and thus less access to alternative measures.

Isolated data reveal a disproportion. However, this emerges in its full magnitude when the sum of discriminations is taken into account. Moreover, this disadvantage does not merely affect those entering the penal system but produces intergenerational effects. Indeed, the fact that many young men from black communities enter prison creates a symbiosis between the social life of communities and the prison environment (Wacquant 2001). The incarceration of many fathers contributes to the impoverishment of black and Latino families, creating a vicious circle that, by keeping their children in a condition of social marginality, makes it more difficult for those children to enter the labor market once they grow up and encourages them to enter illegal markets, primarily by drug dealing (Re 2022).

Finally, to complete the picture, it is necessary to mention the data concerning the death penalty: 41% of people on death row in the United States are white, 42% black, and 14% Latino (Death Penalty Information Center 2022; data as of January 2022). This is a very large disproportion when one considers the fact that whites are 61.6% of the population, blacks 12.4%, and Latinos 18.7%[10].

It should be noted that this discriminatory system was consolidated at a time in history when U.S. society and institutions began to condemn the segregationist practices and slavery on which the nation had been built. Such condemnation, however, as Critical Race Theorists have observed (cf. Thomas and Zanetti, 2005; Delgado and Stefancic 2012), only rarely led to the adoption of incisive policies to redress the social, economic, and cultural damage that black communities suffered. Indeed, the principle of color blindness has prevailed, whereby institutions must act in a neutral manner. In reality, not only has racial bias continued to be exhibited more or less consciously by practitioners in the criminal justice and police systems, but "color-blind" policies, ranging from "zero tolerance" to the "war on drugs", have played a decisive role in the criminalization of minority groups (cf. Re 2006; Ghandnoosh 2015; Davis 2003; Farber 2022).

A similar picture is seen in the Brazilian criminal and prison system, where 2 out of 3 inmates are black and black female inmates are 68% of the female pris-

---

10 2020 Census (https://www.census.gov/library/stories/2021/08/improved-race-ethnicity-measures-reveal-united-states-population-much-more-multiracial.html, visited on 7 July 2022).

on population; in a framework in which the latter has increased by a remarkable 567.4% over the last fifteen years. The majority of these imprisonments are for crimes related to drug trafficking, an activity in which the poorest population who do not have access to education and employment are engaged, mainly as drug dealers (Ribeiro 2019). Women in particular commit crimes to cope with the numerous family needs for which they are often solely responsible. As in the United States, even in Brazil, both criminal and police violence are directed in a highly disproportionate way toward the black population (Ribeiro 2019).

In Europe, too, color-blind penal and penitentiary policies and practices often have discriminatory outcomes. Moreover, in most European countries, the only datum recorded is nationality, and racial discrimination is therefore more difficult to identify. Foreign inmates are overrepresented in the prisons of many European countries, particularly in Southern Europe. It should also be emphasized that these "foreigners" come almost exclusively from countries that exhibit emigration to Europe and are therefore to be considered not merely to be "foreigners" but more properly to be "migrants" (on these labels, that as the terms related to "race" have to be interpreted as social and institutional constructs, cf. Quassoli 2021).

The share of prisoners with foreign citizenship varies across EU Member States. The highest proportions in in the biggest EU countries were recorded in Greece (59.8%), Austria (50.6%), Belgium (43.4%), Estonia (33.8%) and Italy (32.5%)[11]. Looking at the official population data of the EU member states, the numbers of foreign residents differ significantly. However, even in countries whose share of foreign residents is higher, this population is still overrepresented in prison. In Italy, for example, foreigners are less than 9% of the resident population[12] but more than 32% of prisoners.

Research pertaining to the social structure of the prison population has also shown that in some European countries, the proportion of citizens of foreign origin among inmates is significant, as are those of Gypsies, Roma and Travelers in countries in which these minorities are present[13]. The overrepresentation of foreigners, citizens of foreign origin and Gypsies, Roma and Travelers in prison is both an outcome of the structural discrimination to which members of these categories are subject in European societies and a reason for their further stigmati-

---

11 Data referred to 2020, cf. https://ec.europa.eu/eurostat/statistics-explained/index.php?title= Prison_statistics#in_5_prisoners_had_a_foreign_citizenship_in_the_reporting_country, visited on 7 July 2022.
12 Cf. http://stra-dati.istat.it/, visited on 7 July 2022.
13 For Eastern and Central European states, cf., e.g., Cace and Lazar 2003; for the UK, cf. The Traveller Movement 2021.

zation. Indeed, common sense interprets this as evidence of their "tendency to delinquency" (e.g. Melossi 2002; Fiorita et al. 2010–2011; Jobard 2006). As already discussed with respect to the United States, however, here too the overrepresentation of these groups in prison stems primarily from the conditions of social marginality and urban segregation in which many of these people live. This, on the one hand, may facilitate their employment in illegal markets and, on the other, can expose them to more frequent controls than those faced by citizens living in residential or rural areas (FAIRTRIALS 2020).

Both in Europe and in the United States, police stops are frequently based on racial profiling, a technique whose use has increased over the years with the adoption of counterterrorism measures (cf. Center for Human Rights and Global Justice 2006). Like blacks and Latinos in the United States, foreign-born citizens and Gypsies, Roma and Travellers in Europe usually have fewer economic and social resources both to defend themselves in court and to obtain alternative measures. Finally, cases of arbitrary discrimination due to the racist behavior – either conscious or unconscious – of police and/or judges have been reported (cf., e.g., Nwabuzo 2021; Quassoli 1999). Especially in Southern European countries and some Balkan countries whose borders correspond to those of the European Union, however, a significant cause of the overrepresentation of foreigners in prison is the lack of economic and social support as well as migration policies that favor the precarization of the legal status of foreigners, hindering their access to work and housing (Calavita 2005; Palidda 2016; Quassoli 2021).

3 Colonialism's active root

Structural racism is mostly a legacy of the colonial symbolic and material order. The colonial fracture runs throughout the body of contemporary neoliberal democracies. While often denied in public discourse, it continually resurfaces. It was already present in the building of modern Western states; the colony is the negative mirror of "European civilization". To mention only the major European colonial powers, one cannot grasp Britishness without referring to the English colonial empire. Indeed, the myth of the distinguishing characteristics of British civilization was forged during colonialism and was nurtured by studies that, especially, from the development of social Darwinism onward, have validated the prejudice that some races and/or peoples were superior to others. The very idea of an Anglo-Saxon race – which also fuels WASP-related racism in the United States – is rooted in this climate created by the need to justify colonial conquests. In Britain, these racist conceptions merged with social elitism to justify the aristocratic class structure of British society, especially since the British elite was composed of families whose history was intertwined with that of colonial conquest. Racism and elitism were thus welded together. According to some scholars, this combination still inspires many policies adopted in England, e. g., in the field of education, and this sense of superiority and nostalgia for the Empire can be considered to be one of the factors that contributed to Brexit (cf. Dorling and Tomlinson 2019).

Similar remarks can be made about France. While Britishness embodies the British sense of superiority, in France it is rather the idea of *grandeur* that accompanied the development of the *République* and the colonial empire. Here, it was not race myths that played a central role – although France was the home of many "scientific racists" (cf. Todorov 1989; Mosse 1978) and of Arthur de Gobineau (1967 [1853–1855]), author of the "Essai sur l'inégalité des races humaines" – but rather a form of universalism that can be paradoxically called ethnocentric, in the name of which the *citoyen* has always been considered as a member of a special *humanité*.

The notion of "ethnocentric universalism" (Re 2012) builds on the concept of "universalist racism" theorized by Wieviorka (1998). This form of racism differentiates and creates hierarchies among peoples, but it does not propose to keep different "races" separated. Rather, it aims to assimilate peoples that are considered inferior within the great "universal nation". The traits of such a "nation", however, are ethnocentrically identified and reflect those of the colonial power. The "mère-patrie" shows a benevolent face. It can select and authorize the access of non-European peoples to modernity and progress, provided they bend to its will and renounce their "backward" cultures. The "mother-nation" is then an ambiguous figure of a nurturer and educator. She is identified with the fertile and healthy body of the white woman who has primarily nurtured and educated the children she bore but who is ready to help others as well, as long as they passively allow her to guide them[14].

Hierarchies and forms of inferiorization that rest on the history of peoples, cultures and/or the myth of progress can thus coexist with a universal concep-

---

14 A significant analysis of the intertwining of sexism and racism in the construction of the French nation has been proposed by Elsa Dorlin (2009). The "mother-nation" that she portrays is partially different from the one proposed here, since she does not emphasize this figure's universal projection. However, her documented analysis shows how the image of the mother served after the French Revolution as a way to conceive of the nation as consisting of brothers, moving away from the image of the sovereign's body that symbolized the aristocratic conception of the nation under the *Ancien Régime*. Dorlin also stresses the fact this idea of brotherhood helped in identifying the French colonizers as members of the same people while they were away from their homeland.

tion of humanity. Such coexistence was already present in a significant part of European liberal thought (cf. Losurdo 2005) and can also be found today in some rhetoric that sets the West in opposition to "developing countries", i.e., areas of the world that are considered to be "backward" and must therefore "learn" from Western knowledge and experience.

The idea of the healthy, generating and educating mother is also central to the construction of biopolitics and governmentality. As is well known, both terms were coined by Michel Foucault in his studies concerning the functioning of power in modern European states. The first refers to a form of politics – and rationality – that aims "to ensure, sustain, and multiply life, to put this life in order" (Foucault 1976, p. 138). This coincides with the modern state's need to produce and regulate its own population. The second refers to the different social and institutional mechanisms and practices through which the population is managed and controlled in order to foster its prosperity, health, productivity, etc. The colonial fracture has played a crucial role in the refinement of such governmental techniques of population management. Indeed, colonial empires, on the one hand, needed special care for the European citizens who went to the colonies and risked their health and lives to forge the new colonial society enriching the motherland and, on the other hand, were required to manage their relationship with the colonized in terms of assimilation or separation.

Foucault himself stressed the relationship between biopolitics and state's racism, coming to consider racism as the essential characteristic of the modern biopolitical state. Biopolitical population management thus includes genocide and ethnocide, as becomes very clear when we study colonial history and the narrative that it produced (Re 2012). The modern power that shapes the population and governs life retains a residuum of the power to give death, which characterized sovereignty under the *Ancien Régime*. This is regulated precisely by racism – whether "scientific" or "universalist" – which justifies the distinction between those whose lives must be cared for and protected and those who can be killed or left to die. As Foucault (1976; 2011) has argued, racism is thus a technology of power.

Aimé Césaire (2001, p. 36) spoke of the "boomerang effect" of colonization. Objectification devices created in colonial times to enable the brutalization of colonized populations were later used within European territory as well. Totalitarianisms have in fact "applied to Europe colonialist procedures which until then had been reserved exclusively for the Arabs of Algeria, the 'coolies' of India and the 'niggers' of Africa" (Césaire 2001, p. 36). However, the history of colonialism – which intertwined with the development of racism – not only affected the totalitarianisms of the 20th century. It primarily accompanied the development of European modern states, whose political and legal order was founded on respect for the civil and political rights of citizens and the spread of political liberalism (Losurdo 2005; Pitts 2005; Re 2012).

Racism is therefore not alien to liberal democracies. It cannot be regarded as an accidental pathology. As Achille Mbembe (2016a) has argued, liberal democracy has a "solar body" and a "nocturnal body", a remnant of its violent genesis, which is disguised by the original myths of isonomy and self-government. The colonial system, the slave system, and the patriarchal system contributed greatly to the development of modern capitalist societies, leaving this "bitter deposit". Liberal democracies, born as "communities of separation", were based on the distinction between those who were within the "sacred space" that defined the "community of the free individuals" (Losurdo 2005, p. 305), understood as "chosen people", and those who were destined to remain outside: the "non-free" (i.e. slaves, women, the poor, prisoners, the mentally ill, the disabled, homosexuals, etc.), who were present within the territory of the state but excluded from the social pact, and the "barbarians" (i.e. the colonized), who were kept outside the boundaries of the metropolis or destined for ethnocide via violent assimilation. This mechanism of separation and/or violent assimilation has never been removed, although the political and social conflicts that have marked modernity have gradually admitted new categories of people to the "sacred space". It is a division that influences the device of democratic citizenship and thus facilitates the institutionalization of a regime of inequality on a planetary scale (Mbembe 2016a).

More than a "bitter deposit", colonialism can then be described as an active root, since the regime of inequality that it contributed to build deeply characterizes contemporary power relations. Indeed, we live in a profoundly unequal world. The 2022 "World Inequality Report" (Chancel et al. 2021) notes that the richest 10% of the world's population receives 52% of global income, while the poorest half receives only 8.5%. Global wealth is even more inequitably distributed, with the poorest half of the world's population owning just 2%, while the richest 10% own 76%. This picture shows the steady growth – which also occurred during the COVID-19 pandemic – in the assets of a few "super-rich" (520,000 adults, mostly men), who hold more than 3% of global wealth.

Data also show the persistence of imbalances inherited from colonialism. The Middle East and North Africa region is where inequalities are strongest. Again, inequalities have decreased when comparing countries but have also increased within them, as, especially in the poorest countries, the benefits of globalization have generally been the privilege of narrow social minorities, often elites who ensure the maintenance of power structures and geopolitical relations inherited from colonialism. Finally, the road to gender equality remains long: women receive less than 35% of labor income globally (a percentage that has increased by only 4% since 1990). Since inequality and discrimination are often intersectional, such a gender gap is especially suffered by women who are members of minority groups. Discrimination against women depends greatly on their commitment to care labor, and responsibility for the family. The Report "Time to Care", which was published by Oxfam International immediately prior to the spread of COVID-19, highlighted that care work is mainly carried out by women worldwide. It is often unpaid work. When it is paid, care workers receive very low wages:

> Across the world unpaid and underpaid care work is disproportionately done by poor women and girls, especially those from groups who, as well as gender discrimination, experience discrimination based on race, ethnicity, nationality, sexuality and caste. Evidence also shows that inequalities and discrimination based on ethnicity, caste and race leave certain groups with heavier unpaid care responsibilities, and lower incomes, than others. (Lawson et al. 2020, p. 13)

In further detail:

> of the estimated 67 million domestic workers worldwide, 80% are women. Most are women from marginalized groups, who face discrimination based on sex, race, ethnicity, class and caste. Many are driven into domestic work in wealthier countries due to high levels of poverty and exclusion in their own countries. This has created global care chains where care is transferred across countries, from high income women to lower income women. These care chains are sometimes encouraged by cash-strapped governments hoping to raise revenues through remittances. Others are forced or trafficked into domestic work where they are unpaid, controlled or even imprisoned by employers. It is estimated that the 3.4 million domestic workers in forced labour worldwide are being robbed of $8bn every year, having on average been deprived of 60% of their due wages. (Lawson et al. 2020, pp. 36–37)

Global care chains (cf. Yeats 2009) not only represent a form of exploitation that often reproduces the imbalances created by colonialism but also deprive entire communities of care and love, leaving behind children and the elderly (cf., e.g., Parrenas 2010). Indeed, women who migrate to perform care work often cannot take their children or elderly parents with them. They remain in their own countries, deprived of the affection of their nearest and dearest with significant psychological damage (Soros Foundation Romania 2007).

More generally, current data confirm what inequality studies have been observing for some time, namely, the progressive gap between higher and lower wages, the concentration of wealth, the impoverishment of states and the concomitant increase in private wealth. This system is not only unequal; it is also responsible for the ecological crisis that endangers the life of our species on Earth and whose consequences are currently suffered disproportionally by the poorest people. Recall that "the wealthiest 1% of humanity are responsible for twice as many emissions as the poorest 50%" (Ahmed et al. 2022, p. 34), but it is the poorest people who have contributed least to the climate crisis who suffer the most from it. As Oxfam reminded us: "Climate breakdown kills in a variety of ways: malnutrition, diseases, extreme heat, and more intense and more frequent weather-related natural disasters. The vast majority of these deaths occur in low- and middle-income countries, which have contributed relatively little to greenhouse gas emissions" (Ahmed et al. 2022, pp. 34–35). Such deaths must then be considered inequality-related. In addition, millions of people have to leave their homes because of disasters caused by weather. It was estimated that "more than 94% of the 23.7 million new disaster displacements in 2021 were the result of weather-related hazards such as storms, floods and droughts and were recorded in East Asia and the Pacific and South Asia". However, those people were not the most vulnerable, since the population most vulnerable to climate change includes so-called "trapped populations", who are unable to migrate and suffer most from the consequences of environmental disasters (Migration Data Portal 2022).

Even the pandemic crisis depends on this unequal structure of power. Millions of people died because a significant part of the world's population did not have timely access to healthcare and vaccines. However, the demand to lift vaccine patents by some states and NGOs in several countries has fallen on deaf ears. Oxfam has appropriately called this regime "vaccine apartheid" (Ahmed et al. 2022, p. 11). Suffice to say that more than 80% of the available vaccines have gone to G20 countries, while less than 1% have reached low-income countries (Ahmed et al. 2022, p. 28). Worldwide, the pandemic has hit racialized groups the hardest, especially in Brazil, India and the United States. To mention only some impressive data recalled by the Oxfam Report 2022: "in Brazil, Black people are 1.5 times more likely to die from COVID-19 than White people. (...) In the USA, Native American, Latinx, and Black people have been two to three times more likely than White people to die from COVID-19" (Ahmed et al. 2022, p. 28); "3.4 million Black Americans would be alive today if their life expectancy was the same as White's people's. Before COVID-19, that alarming number was already 2.1 million" (Ahmed et al. 2022, p. 7). Finally, "during the second wave of the pandemic in England, people of Bangladeshi origin were five times more likely to die from COVID-19 compared with the white British population" (Ahmed et al. 2022, p. 9).

Hence, the neoliberal political project that has accompanied the development of "hyperglobalization" (Rodrik 2011) is based on competitive individualism and preaches free competition but favors the concentration of economic power and the strengthening of social hierarchies, including racial ones. Colonialism, with its faithful ally racism, is thus a root that is still alive. Today, not only has the colonial divide not been overcome, it has even been deepened by the dynamics produced by neoliberal globalization.

4 Ongoing dehumanization

For Mbembe (2016a; 2016b; 2020), in the current phase, alongside and increasingly even in place of "biopolitics", we are witnessing the development of "necropolitics", that is, a politics whose goal is to expose an increasingly large part of the population to violence and death in order to preserve the economic and social hierarchies of the capitalist system. This shift can be read in part as a return to the original form of capitalism. If this was in fact animated by the drive to manufacture races and species from the beginning, to calculate everything and convert everything into commodities that can be exchanged, to monopolize the manufacture of the living as such, the resistance it encountered during the 19th and 20th century forced it to temper these drives and preserve, with varying outcomes, a number of fundamental separations without which the end of humanity would have become a clear possibility. On the basis of such separations, a subject is not an object; one cannot arithmetically calculate, sell, and buy everything; not everything is exploitable and replaceable; a certain number of perverse fantasies must be sublimated if they are not to lead to the total destruction of the social (Mbembe 2016a).

However, according to the process of liberation from the constraints imposed on it in late modernity that has already been noted by Streeck (2014) with regard to the relationship between state and market, neoliberal capitalism tends to overcome these limits, to the point of blurring the distinctions between human person, object, animal, machine, thus expanding the model of the management of colonized populations to planetary level. It even challenges the taboo of the fabrication of species and subspecies within humanity itself. Such transcending of boundaries does not originate in a critique of anthropocentrism but is rather a way of reviving ancient forms of dehumanization, aimed at the inferiorization and domination of certain social groups. As Chiara Volpato (2014) recalled:

> To dehumanize means to deny the humanity of the other – individual or group – by introducing an asymmetry between those who enjoy the prototypical qualities of the human and those who are considered lacking or deficient in them. Dehumanization is multifaceted, multiform, flexible. It adapts to places, people, relationships, taking on from time to time the content required by the cultural climate of the moment.

The goal of dehumanization is to authorize violence by denying the moral dignity of the other. It also fosters psychological estrangement between the dominant and the dominated and reassures the former that their fate can never be common to that of the latter. Yet it is precisely this certainty that is increasingly challenged today by the processes of precarization, impoverishment, and increased inequality that are taking place even within Western democracies. If the "Negro" was the excluded in the colonial racist regime, on the basis of a fixed, blood-bound identity, today, this condition extends to larger and larger groups of people. Djamila Ribeiro (2022, p. 13) has pointed out that the quality of "Negro" has historically been – and often still is today, even in individual experience – a violent hetero-ascription. This category was in fact created as part of a process of discrimination, which aimed to legitimize the treatment of human beings, slaves, as commodities. Thus, the creation of the "Negro" was accompanied by a process of erasing the cultures, languages, religious traditions, etc., of "inferior people". This process was met with resistance, but this does not mean that it can be considered to be less violent on both the material and the symbolic levels.

Once again, it should be noted that this colonial root is only seemingly remote. In the context of contemporary neoliberal capitalism, it is in fact continually reactivated. The condition reserved for the "Negro" in the colony is extended to larger and larger portions of humanity. However, this process does not reduce racism. On the contrary, it inspires increasingly explicit forms of racism in part of the population that perceives itself as "White" and fears losing its economic and social well-being. These are motivated by fear of losing the racial privilege institutionalized in the formation of modern Western states and therefore of being united with those who from the beginning have not been fully part of the social pact.

These fears are fueled by the media and political forces that insist on the issue of immigration and/or minority deviance and the equation of immigration/minorities with crime. Fears are fomented only to be periodically contained via seemingly strong policies that are designed to reaffirm the state's ability to 'defend its citizens'. This phenomenon highlights the governing through crime – we might say more appropriately governing through fear – that has been studied in the United States by Jonathan Simon (2007) and that has been considered by Zygmunt Bauman (1998) to be a system of governance that is characteristic of globalized societies, in which nation-states are weak in the face of market forces and merely reassert their sovereign power over marginals and migrants. This mechanism allows conflict to be shifted from economic-social terrain to cultural, religious and even "racial" terrain, effectively fostering the rise of inequality and the expansion of neoliberal rationality. In this framework, racism in fact becomes once again able to spread inequality and strengthen the social hierarchies produced by capitalism. The elites pretend to condemn it in its most explicit manifestations but in fact leave the field open to its unfolding, both because they maintain the racist structures that allow the marginalization of the social groups under attack and because they do not counteract the social insecurity that triggers racist prejudices. Note: to make this argument is not to justify racism nor to view manifestations of racist nationalism or supremacism merely as demands of social justice that have not been satisfactorily answered. Rather, these manifestations, as Giorgia Serughetti (2021) has argued, comply with the neoliberal ideology that nurtures an authoritarian individualism, shaping what Mbembe (2016a; 2016b) has called "the society of enmity".

In fact, the social anger that nurtures racist behaviors does not usually come from the more marginal social classes, but rather from the middle class that fears losing its position. When people from the working class adopt racist prejudices, these views are often more directly related to racist propaganda than personal experience (cf., e.g., Dorling and Tomlinson 2019; Bertuzzi et al. 2019). This anger arises primarily from the refusal to consider or lack of faith in a different social order, that is, one in which it is possible to assert egalitarian instances. The resulting manifestations of violence comply with the violent system of neoliberal structural racism and support the exploitation of discriminated groups. The suffering of these groups compensates the former "privileged" for material impoverishment, whether real or feared. Through the demand to deny rights to minorities – usually to racial, religious, cultural, and sexual minorities alike – an attempt is made to protect the patriarchal and colonial privilege, to maintain a symbolic order: 'if I cannot aspire to a secure social position, let alone a better one, I can at least hope to keep those beneath me at a distance by exerting a violent power over them'[15].

"Pocket-knife" (Mbembe 2016b, p. 31), even goliardic racism, is thus becoming less and less concealed, and the way is opened for the trivialization of racism, for a "narcotic brand of prejudice based on skin colour and expressing itself in seemingly anodyne everyday gestures" (Mbembe 2016b, p. 31). This "nanoracism" (Mbembe 2016a; 2016b) legitimizes exploitation and makes a part of humanity feel "undesirable", "surplus" (Mbembe 2016b, p. 23). It is "the obligatory complement to hydraulic racism", that is, to a subtler racism, which is formed by "micro- and macro-juridical, bureaucratic and institutional apparatuses" (Mbembe 2016b, p. 32) of marginalization and segregation.

---

15 This same reading has been endorsed by some feminist scholars studying contemporary gender-based violence (cf. Pitch 2008; Zeynep 2015; Re et al. 2019).

5 "Unreal lives"

Not only do racial hierarchies subsist, but as Judith Butler (2004) highlighted in the aftermath of September 11th, 2001, the hierarchy between "grievable" and "ungrievable" lives has become increasingly clear.[16] This hierarchy is largely superimposable on the one indicated by Mbembe (2016a) and on those that can be detected in colonial history. This "differential allocation of mourning" operates "to produce and maintain certain exclusive conceptions of the human in a normative sense" (Butler 2004, p. XIV). It designates "what counts as a life to live and a death to grieve over" (Butler 2004, p. XV).

Mbembe (2016a) emphasizes how the "Negro" can be understood as the "Other" of the man-flux, digital, who is infiltrated everywhere by synthetic organs and artificial prostheses of all sorts, of a computer humanity, a new figure of the species, so typical of the new era of capitalism, in which self-reification represents the best opportunity for the capitalization of the self. In both cases, therefore, it is a question of bodies made objects or, more precisely, of "unreal lives", de-realized, on which the exercise of violence no longer causes a scandal. Butler (2004; 2016) has spoken in this regard of "spectral lives", of "zombies" who have already died while remaining alive and that for this reason, once killed, cannot be mourned. These are the same zombies that inhabit our fears, popular movies, and videogames intended for the general public, and they serve as a metaphor precisely for a loss of humanity, not only of the "others" that we have dehumanized, but also of "us", who are reduced to "two-dimensional economic beings" (Mason 2019) besieged by technology.

Mark Levine (2020), based on Mbembe's analysis, has spoken of "necroliberalism", a system rooted in colonial dispossession and capitalism's original "necropolitics", but which today "is justified by neoliberal economic orthodoxies, the 'War on Terror', and growing direct calls for 'pure' identities, racially, ethnically, culturally, and religiously". "Necroliberal governance", Levine further writes, "requires a deeper institutionalization of corruption in political and financial systems and, because of this, an even greater amount of wealth directed to what has been called in the jargon the '1 percent'".

In *Brutalisme*, Mbembe (2020) emphasized that a necessary consequence of "necropolitics" is the devastation of the planet. The predatory and extractive logic proper to the colony has in fact extended to the entire Earth. The "universalization of the Negro condition" (Mbembe 2020), in the era of Capitalocene (Moore, Ed., 2016), in which the ecology of the planet is determined by capital

---

16 In this paragraph and the next, I take up some of the arguments developed in Re 2020.

ism, involves humans and other living beings but does not reduce hierarchies; indeed, it emphasizes them. In a horizon in which, as observed above, the production of "wasted lives" (Bauman 2004) is increasingly an explicit goal of the economic and political system, violent death is exposed with increasing frequency. As in the colony, it becomes routine.

Alexis de Tocqueville (1951–1998, pp. 226–227) wrote, with reference to the French colony of Algeria, that the death of unarmed civilians, including women and children, is part of the "unpleasant necessities" (cf. Tocqueville 1951–1998, pp. 226–227) of colonial war, a war that escapes the "mise en forme" and that inevitably must be fought outside the rules established by the "jus in bello"[17]. Today this state of war has become molecular (Mbembe 2020). The "line" between formal and informal wars is continually crossed. Contemporary wars are in fact not only those that are expressly called: wars that often have their origin in the geopolitical order inherited from colonialism and which are also the subject of greater or lesser attention by Western public opinion based on the nationality, skin color and professed religion of the victims – emblematic in this regard is the favor with which refugees of Ukrainian nationality were received in Europe in spring 2022 compared to the rejection and abuse suffered by refugees from Africa and Asia. Alongside these terrible wars are other informal ones that are fought daily in the slums and inner-cities worldwide, where the inhabitants are squeezed between the violence of crime and that of the police. Moreover, in rich countries, this kind of wars are fought against migrants who are rejected at the borders and interned in camps or exploited in the fields according to a system that recalls the colonial system of "plantations" (cf. Mbembe 2016a; Davis 1981). They are also fought against the children of immigrants and young people belonging to minorities, who are excluded from the full enjoyment of citizenship rights and often criminalized (cf. Bhabha 2014). Finally, the "war against women" (Segato 2016) is fought in every part of the globe and aggravated by the combination of different forms of intersectional discrimination.

---

17 This expression refers to the body of the law of war developed in modern Europe to regulate warfare and protect civilian populations. Such law is independent from the justification for war (*jus ad bellum*), which was at the center of legal, political and ethical speculation in the premodern world and has regained importance since the second half of the 19th century. Nowadays the *jus in bello* refers to international humanitarian law, which governs the way in which warfare is conducted, seeking to limit the suffering caused by the conflict (cf. http://www.icrc.org/en/war-and-law/ihl-other-legal-regmies/jus-in-bello-jus-ad-bellum, visited on 7 July 2022).

6 Necroliberal violence

The deaths that occur within what can be read, in light of Mbembe's reflections (2020), as a continuum of violence, are not only, as mentioned, increasingly normalized; they also call into question the role of states, including those that define themselves as constitutional states: "with brutalism, killing ceases to be an exception. The transposition of the state of war within a civilian state leads to the normalization of extreme situations. The state sets about committing common crimes against civilians" (Mbembe 2020).

Of course, the multiplication of "exceptions", if not outright suspensions of the rule of law, while common to different realities, functions differently where the state appears on the whole to be involved in the killings through the direct participation of its apparatuses or the guarantee of impunity to private powers that perpetrate them and where, on the contrary, some bodies still react to human rights violations thanks to the articulated structure of constitutional states.

Tocqueville suggested to clearly distinguish in the colony between the liberal regime to be applied to European settlers and the despotic regime to be reserved for the natives. In his colonial writings, he insists on this point even if he fears that the authoritarian methods employed by the military and the suppression of the division of powers in the colony could end up infecting the motherland, a fear that has proven to be historically well-founded.

Today, the multiplication of colonial statuses is evident even in constitutional states. It is often enshrined in law, and it represents a serious danger to the stability of (inter)national constitutionalism. This multiplication is becoming increasingly evident in European migration policies, to the point of characterizing EU citizenship itself as an "exclusive citizenship" (Santoro 2007) based on selection rather than inclusion. Such a citizenship can increasingly be obtained by resident foreigners only if they prove to have an adequate level of income and after they go through a long process that involves passage through different statuses, from irregular foreigners to temporarily accepted residents to long-term residents according to paths that are rarely linear and that always feature the risk of returning to the starting point. As Encarnación La Spina (2018, p. 328) has written, starting with the categories of "citizen" and "immigrant", which are assumed to be antagonistic, a series of intermediate categories is consolidated "and semi-citizens appear, made up of those immigrants who have acquired the status of residents but are not citizens, as well as those who have not regularized their administrative situation, thus remaining on the margin of the benefits of residency and, therefore, non-subjects".

It can be argued that restrictive regulations on immigration and access to citizenship aim at a precarization of the legal situation of immigrants, according to a process that continuously oscillates between the two extremes of social inclusion and exclusion. This very rarely allows immigrants to overcome the "iron cage" (La Spina 2018, p. 338). Non-citizens and semi-citizens cannot in fact benefit – *de jure* or *de facto* – from many fundamental rights, in particular political rights and some social rights. In this framework, the distinction between citizens and foreigners and the enucleation of subcategories of immigrants are functional with respect to the maintenance of a relationship of force linking individuals, groups and the State. The path of access to citizenship is therefore studded with obstacles that transform it from a presupposition of inclusion through rights, according to Thomas Marshall's well-known theory (2014 [1950]), into a prize for the chosen few. This filtering system, moreover, operates on the basis of more or less explicit forms of racialization, for which the degree of control to which one is subjected and the possibility of obtaining more easily long-term residence permits, family reunification and, ultimately, citizenship depend on nationality according to a logic that, for example in Italy, privileges communities in which the majority of people profess the Christian religion, such as Filipino or Peruvian communities. As Quassoli (2021, p. 179) recalls, Schengen blacklists and whitelists, which distinguish between third countries whose citizens need a visa from countries whose citizens do not need a visa to enter the area, are based on definite features: in the countries included in the blacklist, the majority of the population is poor, Muslim and "nonwhite".

The obstacles that these people face when entering Europe are, at the same time, a result and a symbol of their undesirability, which is continually reiterated by their detention on the territory and/or their deportation (cf. De Genova 2013). Such a management of "migration" demonstrates that mobility is a privilege and reaffirms a racial hierarchy rooted in colonialism (Quassoli 2021, p. 179). Such a hierarchy is based on differentiated legal statuses, a sort of "globalization of the code de l'indigénat". It is a profoundly violent system of management of racialized populations in both the South and the North of the world. Mbembe and Bauman both emphasize the immobilization of large strata of the world's population. According to this reading, neoliberal globalization is based on a hierarchy of mobility between "tourists" and "vagabonds" (Bauman 1998), between citizens who are free to move on a planetary level and "glebae adscripti", human beings confined to the lands dominated by their masters. For the "racialized fractions of post-industrial societies", "mobility choices are often limited to ghetto assignment or incarceration" (Mbembe 2020). In particular, those from former colonies who try to escape this system of immobilization through emigration become "public enemies" and are treated as "prey" according to the logic of hunting that has always been applied to groups that have been inferiorized even through assimilation to the animal world: black slaves, Native Americans, colonial peoples, Jews, Roma, women, homosexuals, etc.

Those who try to cross borders become the border themselves, their "migrant" body becomes in fact the place of control and repression. From this point of view, the borders of Europe are mobile and fragmented. They multiply into thousands of "border-bodies" (Mbembe 2020). They are, above all, bodies of African men, women and children who, more than other foreigners, are considered threats: "In other words, it is the body of the African, of each African taken individually and of all Africans as a racialized class, that now constitutes the frontier of Europe. It is, therefore, a mobile, itinerant frontier, determined no longer by fixed lines, but by bodies in motion" (Mbembe 2020). It is this "border-body" that "is forbidden to host or protect (...), to save from drowning on the high seas or from dehydration in the middle of the desert" (Mbembe 2020). And so "violence at and across borders has become one of the defining features of the contemporary condition. Little by little, the fight against the so-called illegal migrations takes the form of a social war now conducted on a planetary scale" (Mbembe 2020). The objective of this "hunt", however, does not seem to be so much, as Mbembe seems to think, annihilation as an end in itself as the selection and the precarization of the existences of those who manage to cross the borders, an "inclusion through illegalization" (De Genova 2005, p. 234), which allows the achievement of forms of "differential inclusion" (Mezzadra 2013, p. 423).

The "wasted" and "ungrievable" lives – an expression that cannot fail to recall in Italy the image of the coffins of the 368 migrants who were shipwrecked on October 3, 2013 and lined up in a bare room in Lampedusa, without anyone present to commemorate them – are a warning for those who have managed to expatriate and who remain the potential "prey" of a system that can reject them at any time. In this sense, once again, "the universalization of the Negro condition" (Mbembe 2020) refers to a generalized decline in protection, to a precarization of lives, in the sense clarified by Butler (2004).

7 Conclusion

Faced with this continuous reproduction of hierarchies, the "anti-racist" discourse in our democracies is weak and can even become hypocritical. If it is not combined with a strong condemnation of violence and the revival of a universalism that must be "taken seriously", it can contribute to the fiction of an inclusive society and the reestablishment of the neoliberal principle of individual competition obfuscate the social hierarchies and the obstacles that many people face every day.

The battle against racism is not only educational and cultural, although education and culture are fundamental tools. It must also be fought at a deeper level to break down the racist structures we have traced here and to reaffirm the "political project of equality". In many Western democracies – as in Italy – the political discourse contrasts "do-gooders", who are animated by humanitarian ideals and often perceived as privileged individuals whose well-being is not threatened by globalization, and "realists", who set the need to protect fundamental rights in opposition to the need to limit immigration and/or crime. This false opposition distorts the perception of reality. In fact, the so-called "do-gooders" are often nothing more than people fighting to preserve the structure of constitutional states, the guarantees of *habeas corpus*, and the democratic foundation on which they rest. Accepting the necropolitical hierarchies means in fact burying the political project of (inter)national constitutionalism and its dream of equality, with a devastating impact on the rights of all.

In turn, the "realists" think that immigration and/or crime can be limited by progressively increasing the violence of the system. There is nothing realist about policies that are based on violent border management and/or the hypertrophy of the criminal justice system. First, because these policies are often based on an exaggerated alarm regarding problems of different nature that need differentiated solutions. Secondly, because facing the problems that necessarily arise in a global society realistically would require finding solutions – which in part already exist, including humanitarian corridors, family reunification, sponsor mechanisms and asylum policies, social policies, and alternative measures – to manage social inclusion in a differentiated way while protecting human rights and fighting mafias and corruption.

In this scenario, we therefore need to preserve the values affirmed throughout the second half of the 20th century while combining their preservation with a transformation that allows us to adapt them to the postmodern condition in which we live. On the foundations of (inter)national constitutionalism, it is possible to establish a constitutional state that can guarantee the rights of non-paradigmatic subjects, including racial minorities, making their voices heard. The critique of racism must therefore join forces with the more general critique of neoliberal violence and lead us to outline alternative paradigms. These, as mentioned above, do not have to arise *ex nihilo* but can take root in the political and legal system forged by (inter)national constitutionalism based on the principle of substantive equality and the effectiveness of social rights through adequate welfare measures. On this basis, we can develop new forms of bottom-up universalism, starting by listening to different subjectivities and including those that have been so far invisibilized (Zanetti 2019). The struggle against racism is thus not merely a struggle against neoliberalism but also a struggle for law, which aims to transform our "careless states" (Care Collective 2020) and our "societies of enmity" into "caring democracies" (Tronto 2013) and caring communities (Care Collective 2020) that are open to pluralism and kept together by bonds of solidarity. This is a struggle that must necessarily combine the activism of social movements with the commitment of institutions for the purpose of rebuilding a plural and inclusive "public sphere".

References

Ahmed, Nabil; et al. (2022): "Inequality Kills. The unparalleled action needed to combat unprecedented inequality in the wake of COVID-19". Oxford: Oxfam International. https://www.oxfam.org/en/research/inequality-kills, visited on 6 July 2022.

Baratta, Alessandro. (1980): *Introduzione alla sociologia giuridico-penale*. Bologna: Lorenzini.

Bauman, Zygmunt. (1998): *Globalization. The Human Consequences*. Cambridge, Oxford: Polity Press-Blackwell.

Bauman, Zygmunt. (2004): *Wasted lives. Modernity and its Outcasts*. Cambridge: Polity Press.

Bertuzzi, Niccolò; Caciagli, Carlotta; Carusi, Loris. (2019): *Popolo chi? Classi popolari, periferie e politica in Italia*. Roma: Futura editrice.

Bhabha, Jacqueline. (2014): *Child Migration and Human Rights in A Global Age*. Princeton: Princeton University Press, ebook.

Bourdieu, Pierre. (1980): *Le sens pratique*. Paris: Editions de minuit.

Brown, Wendy. (2015): *Undoing the demos. Neoliberalism's Stealth Revolution*. New York: Zone books, ebook.

Butler, Judith. (2004): *Precarious Life. The Powers of Mourning and Violence*. London: Verso.

Butler, Judith. (2016 [2009]): *Frames of War. When is Life Grievable?*. London: Verso, ebook.

Cace, Sorin; Lazar, Cristian. (2003): "Discrimination against Roma in Criminal Justice and Prison Systems in Romania. Comparative Perspective of the Countries in Eastern and Central Europe". Bucharest: Penal Reform International Romania. https://www.publicsafety.gc.ca/cnt/rsrcs/lbrr/ctlg/dtls-en.aspx?d=PS&i=15654956, visited on 6 July 2022.

Calavita, Kitty. (2005): *Immigrants at the Margins. Law, Race, and Exclusion in Southern Europe*. Cambridge: Cambridge University Press.

Casalini, Brunella. (2007): "Cittadinanza e 'riproduzione' della razza negli Stati Uniti d'America". In: Re, Lucia (Ed.): *Differenza razziale, discriminazione e razzismo nelle società multiculturali*. Vol. 2. Reggio Emilia: Diabasis, pp. 125–134.

Center for Human Rights and Global Justice. (2006): "Irreversible Consequences. Racial Profiling and Lethal Force in the 'War on Terror'. Briefing Paper". New York: New York University School of Law. http://www2.ohchr.org, visited on 27 June 2022.

Césaire, Aimé. (2001): *Discourse on Colonialism*. New York: Monthly Review Press.

Chancel, Lucas; et al. (Eds.) (2021): "World Inequality Report". Paris: World Inequality Lab. https://wir2022.wid.world/insights/, visited on 7 July 2022.

Clavero, Bartolomé. (2002): "Stato di diritto, diritti collettivi e presenza indigena in America". In: Costa, Pietro; Zolo, Danilo (Eds.): *Lo Stato di diritto. Storia, teoria, critica.* Milano: Feltrinelli, pp. 537–565.

Congresso Nacional. (1993): "Relatório N° 2 de 1993". Brasilia. https://www2.senado.leg.br/ bdsf/bitstream/handle/id/85082/CPMIEsterilizacao.pdf?sequence=7&isAllowed=y, visited on 6 July 2022.

Davis, Angela Y. (1981): *Women, Race & Class.* New York: Random House Inc.

Davis, Angela Y. (2003): *Are Prisons Obsolete?.* New York: Seven Stories Press.

Death Penalty Information Center. (2022): "Facts About the Death Penalty". June 24. Washington DC: DPIC. https://documents.deathpenaltyinfo.org/pdf/FactSheet.pdf, visited on 6 July 2022.

De Genova, Nicholas. (2005): *Working the Boundaries: Race, Space, and "Illegality" in Mexican Chicago.* Durham (NC): Duke University Press.

De Genova, Nicholas. (2013): "Spectacles of Migrant 'Illegality': The Scene of Exclusion, the Obscene of Inclusion". In: *Ethnic and Racial Studies* 36. N. 7, pp. 1180–1198.

Delgado, Richard; Stefancic, Jean (Eds.) (2012): *Critical Race Theory. An Introduction.* Second Edition. New York: New York University Press.

Dorlin, Elsa. (2009): *La matrice de la race. Généalogie sexuelle et coloniale de la Nation française.* Paris: La Découverte, ebook.

Dorling, Danny; Tomlinson, Sally (2019): *Rule Britannia. Brexit and the End of Empire.* London: Biteback, ebook.

FAIRTRIALS. (2020): "Un-covering anti-Roma discrimination in criminal justice systems in Europe". https://www.fairtrials.org/sites/default/files/publication_pdf/FT-Roma_report-final.pdf, visited on 6 July 2022.

Farber, David. (Ed.) (2022): *The War on Drugs. A History.* New York: New York University Press, ebook.

Ferrajoli, Luigi. (2018): *Manifesto per l'uguaglianza.* Roma-Bari: Laterza, ebook.

Fiorita, Nicola; Giolo, Orsetta; Re, Lucia (Eds.) (2010–2011): "La minoranza insicura. I Rom e i Sinti in Europa". In: *Jura gentium. Rivista di filosofia del diritto internazionale e della politica globale* 8. Special Issue.

Foucault, Michel. (1975): *Surveiller et punir. Naissance de la prison.* Paris: Gallimard.

Foucault, Michel. (1976): *La volonté de savoir. Histoire de la sexualité.* Vol. 1. Paris: Gallimard.

Foucault, Michel. (2011): *Il faut défendre la société. Cours au Collège de France (1975–1976).* Paris: Gallimard.

Franzina, Emilio; Stella, Gian Antonio (2002): "Brutta gente. Il razzismo anti-italiano". In: Bevilacqua, Piero; De Clementi, Andreina; Franzina, Emilio (Eds.): *Storia dell'emigrazione italiana. Arrivi.* Vol. 2. Roma: Donzelli, pp. 283–312.

Ghandnoosh, Nazgol. (2015): *Black Lives Matter. Eliminating Racial Inequity in the Criminal Justice System.* Washington D.C.: The Sentencing Project. https://www.sentencingproject. org/publications/black-lives-matter-eliminating-racial-inequity-in-the-criminal-justice-system/, visited on 6 July 2022.

Girard, René. (1982): *Le bouc émissaire.* Paris: Grasset.

Gobineau, Arthur, de. (1967): *Essai sur l'inégalité des races humaines* (1853–1855). Paris: P. Belfond.

International Commission of Inquiry on Systemic Racist Police Violence Against People of African Descent in the United States. (2021): "Report of the International Commission of Inquiry on Systemic Racist Police Violence Against People of African Descent in the United States", March 2021. https://inquirycommission.org, visited on 28 June 2022.

Jacobson, Matthew Frye. (1998). *Whiteness of A Different Color. European Immigrants and the Alchemy of Race.* Cambridge. (Mass.), London: Harvard University Press, ebook.

Jobard, Fabien. (2006): "Police, justice et discriminations raciales". In: Fassin, Didier; Fassin Éric (Eds.): *De la question sociale à la question raciale?.* Paris: La Découverte, pp. 211–229.

Kennedy, John Fitzgerald. (1964): *A Nation of Immigrants.* New York: Harper & Row.

King, Desmond. (2002): *Making Americans. Immigration, Race and the Origins of the Diverse Democracy.* Cambridge (Mass.) and London: Harvard University Press, ebook.

La Spina, Encarnación. (2018): "Immigrati nell'Europa meridionale. Quando 'non si nasce ma si diventa' giuridicamente 'particolarmente vulnerabili'?". In: Bernardini, Maria Giulia; Casalini, Brunella; Giolo, Orsetta; Re, Lucia (Eds.): *Vulnerabilità. Etica, politica, diritto.* Roma: IF Press, pp. 315–340.

Lawson, Max et al. (2020): "Time to Care. Unpaid and underpaid care work and the global inequality crisis". Oxford: Oxfam International. https://oxfamilibrary.openrepository.com/bitstream/handle/10546/620928/bp-time-to-care-inequality-200120-en.pdf, visited on 6 July 2022.

Levine, Mark. (2020): "From neoliberalism to necrocapitalism in 20 years". In: *Aljazeera.* https://www.aljazeera.com/indepth/opinion/neoliberalism-necrocapitalism-20-years-200715082702159.html, visited on 7 July 2022.

Losurdo, Domenico. (2005): *Controstoria del liberalismo.* Roma-Bari: Laterza.

Marshall, Thomas Humphrey. (2014 [1950]): "Citizenship and Social Class". In: Marshall, Thomas Humphrey; Bottomore, Tom: *Citizenship and Social Class.* Part I. London: Pluto Press, ebook.

Mason, Paul. (2019): *Clear Bright Future. A Radical Defence of the Human Being.* London: Penguin, ebook.

Mauer, Marc. (1999): *Race to Incarcerate.* New York: The New Press.

Mazzarese, Tecla. (2018): "I migranti e il diritto ad essere diversi nelle società multiculturali delle democrazie costituzionali". In: Cerrina Feroni, Ginevra; Federico, Veronica (Eds.): *Strumenti, percorsi e strategie dell'integrazione nelle società multiculturali.* Napoli: ESI, pp. 63–85.

Mbembe, Achille. (2016a): *Politique de l'inimitié.* Paris: La Découverte.

Mbembe, Achille. (2016b): "The society of enmity". In: *Radical Philosophy* 200, pp. 23–35.

Mbembe Achille. (2020): *Brutalisme.* Paris: La Découverte, ebook.

Melossi, Dario. (2002): *Stato, controllo sociale, devianza.* Milano: Bruno Mondadori.

Mezzadra, Sandro. (2013): "Moltiplicazione dei confini e pratiche di mobilità". In: *Ragion Pratica* 41. N. 2, pp. 413–431.

Migration Data Portal. (2022): "Environmental Migration". https://www.migrationdataportal.org/themes/envornmental_migration_and_statistics. Last update 21 June 2022, visited on 6 July 2022.

Moore, Jason W. (Ed.) (2016): *Anthropocene or Capitalocene? Nature, History and the Crisis of Capitalism.* Oakland: PM Press, ebook.

Mosse, George Lachmann. (1978): *Toward the Final Solution. A history of European racism.* New York: Howard Fertig.

National Research Council. (2014): "The Growth of Incarceration in the United States". Washington D.C.: The National Academies Press. https://nap.nationalacademies.org/cata log/18613/the-growth-of-incarceration-in-the-united-states-exploring-causes, visited on 6 July 2022.

Noiriel, Gérard. (2007): *Immigration, antisémitisme et racisme en France (XIXe-XXe siècle). Discours publics, humiliations privées.* Paris: Fayard.

Nwabuzo, Ojeaku. (2021): "The Sharp Edge of Violence. Police Brutality and Community Resistance of Racialised Groups". Brussels: European Network Against Racism. http:// enar-eu.org, visited on 28 June 2022.

Palidda, Salvatore. (2008): *Mobilità umane. Introduzione alla sociologia delle migrazioni.* Milano: Raffaello Cortina Editore.

Palidda, Salvatore (Ed.) (2016): *Racial Criminalization of Migrants in the 21st Century.* London: Routledge.

Parrenas, Rhacel Salazar. (2010): "Transnational Mothering. A Source of Gender Conflicts in the Family". In: *North Carolina Law Review* 88. N. 5, pp. 1825–1856.

Pitch, Tamar. (2004): *I diritti fondamentali: differenze culturali, disuguaglianze sociali, differenza sessuale,* Torino: Giappichelli.

Pitch, Tamar. (2008): "Qualche riflessione attorno alla violenza maschile contro le donne". In: *Studi sulla questione criminale* 3. N. 2, pp. 7–13.

Prison Policy Initiative. (2020): "Annual Report 2019–2020". October, Northampton (Mass.). https://www.prisonpolicy.org/reports/pie2020.html, visited on 6 July 2022.

Pitts, Jennifer. (2005): *A Turn to Empire: the Rise of Imperial Liberalism in Britain and France.* Princeton: Princeton University Press.

Quassoli, Fabio. (1999): "Immigrazione uguale criminalità: rappresentazioni di senso comune e pratiche degli operatori di diritto". In: *Rassegna italiana di sociologia* 1, pp. 43–76.

Quassoli, Fabio. (2021): *Clandestino. Il governo delle migrazioni nell'Italia contemporanea.* Milano: Meltemi, ebook.

Re, Lucia. (2006): *Carcere e globalizzazione. Il boom penitenziario negli Stati Uniti e in Europa.* Roma-Bari: Laterza.

Re, Lucia. (2012): *Il liberalismo coloniale di Alexis de Tocqueville.* Torino: Giappichelli.

Re, Lucia. (2020): *Democrazie vulnerabili. L'Europa dall'identità alla cura.* Pisa: Pacini.

Re, Lucia. (2022): "Criminalità e criminalizzazione: selettività sociale, discriminazione razziale, diseguaglianza di genere". In: Pitch, Tamar (Ed.): *Devianza e questione criminale. Temi, problemi e prospettive.* Roma: Carocci, pp. 45–62.

Re, Lucia; Rigo, Enrica; Virgilio, Maria (Milli) (2019): "Le violenze maschili contro le donne: complessità del fenomeno ed effettività delle politiche di contrasto". In: *Studi sulla questione criminale* 14. N. 1–2, pp. 9–33.

Ribeiro, Djamila. (2019): *Pequeno Manual Antirracista.* São Paulo: Companhia Das Letras, ebook.

Rigione, Salvatore. (2020): *Sulle tracce di una mitografia italiana della razza nella rincorsa coloniale,* Pisa: ETS.

Roberts, Dorothy. (1997): *Killing the Black Body: Race, Reproduction and the Meaning of Liberty.* New York: Pantheon.

Rodrik, Dani. (2011): *The Globalization Paradox. Democracy and the future of the world economy.* New York: W.W. Norton.

Santoro, Emilio. (2007): "La cittadinanza esclusiva: il carcere nel controllo delle migrazioni". In: Re, Lucia (Ed.): *Differenza razziale, discriminazione e razzismo nelle società multiculturali.* Vol. 2, pp. 44–68. Reggio Emilia: Diabasis.

Segato, Rita Laura. (2016): *La Guerra contra las mujeres.* Madrid: Traficantes de sueños, ebook.

Simon, Jonathan. (2007): *Governing Through Crime. How the War on Crime Transformed American Democracy and Created a Culture of Fear.* Oxford, New York: Oxford University Press.

Soros Foundation Romania. (2007): "The effects of migration: the children left behind". https://ec.europa.eu/migrant-integration/library-document/effects-migration-children-left-behind_en, visited on 6 July 2022.

Streeck, Wolfgang. (2014): *Buying Time. The Delayed Crisis of Democratic Capitalism.* London: Verso.

The Care Collective. (2020): *The Care Manifesto. The Politics of Interdependence.* London: Verso, ebook.

The Traveller Movement. (2021): "Gypsy, Roma and Traveller Women in Prison". London: The Traveller Movement. https://wp-main.travellermovement.org.uk/wp-content/uploads/2021/09/GRT-Women-in-Prison-Report-2021.pdf, visited on 6 July 2022.

Thomas, Kendall; Zanetti, Gianfrancesco (Eds.) (2005): *Legge, razza, diritti.* Reggio Emilia: Diabasis.

Tocqueville, Alexis de. (1951–1998): *Œuvres Complètes.* Vol. III, 1. Paris: Gallimard.

Todorov, Tzvetan. (1989): *Nous et les autres. La réflexion française sur la diversité humaine.* Paris: Seuil.

Tronto, Joan. (2013): *Caring Democracy. Markets, equality and justice.* New York, London: New York University Press.

U.S. Department of Justice. (2020a): "Prisoners in 2019". October. NCJ 255115. https://bjs.ojp.gov/content/pub/pdf/p19.pdf, visited on 6 July 2022.

U.S. Department of Justice. (2020b): "Contacts Between Police and the Public, 2018 – Statistical Tables". December. NCJ 255730. https://bjs.ojp.gov/content/pub/pdf/cbpp18st.pdf, visited on 6 July 2022.

U.S. Department of Justice. (2021): "Prisoners in 2020- Statistical Tables". December. NCJ 302776. https://bjs.ojp.gov/content/pub/pdf/p20st.pdf, visited on 6 July 2022.

Volpato, Chiara. (2014): *Deumanizzazione. Come si legittima la violenza.* Roma-Bari: Laterza, ebook.

Wacquant, Loïc. (2001): "Deadly Symbiosis: When Getto and Prison Meet and Mesh". In: *Punishment and Society* 3. No. 1, pp. 95–133.

Wieviorka, Michel. (1998): *Le racisme. Une introduction.* Paris: La Découverte.

Yeates, Nicola. (2009): *Globalizing Care Economies and Migrant Workers. Explorations in global care chains.* London: Palgrave, Macmillan.

Zanetti, Gianfrancesco. (2019): *Filosofia della vulnerabilità. Percezione, discriminazione, diritto,* Roma: Carocci.

Zeynep, Direk. (2015): "Una risposta femminista a *Violence et civilité*". In: *Jura gentium. Rivista di filosofia del diritto internazionale e della politica globale* 12. Pp. 167–80.

Zolo, Danilo. (2000): *Chi dice umanità. Guerra, diritto e ordine globale.* Torino: Einaudi.

Carlo Botrugno
The Racial Epistemicide of Bioethics

Abstract: The killing of George Floyd on May 25th, 2020, led to long-lasting pro-
tests and triggered a strong reaction and global social condemnation. However,
racism can take several forms besides brutal violence and killing. Unveiling and
fighting hidden racism is a major goal of the medical, healthcare, and bioethics
fields. Bioethics in particular must take action and contribute to reversing the
direction of our racist societies, promoting equality and respect, thus defending
everyone's right to exist. To achieve this purpose, as argued in this work, there is
a need to intervene at the roots of racism in bioethics, i.e., the production of
knowledge relevant to our discipline and related subject areas. It is indeed
this epistemic ground that informs the emergence of policies, practices, views,
and perspectives (both racist and anti-racist). In this chapter, I focus on "racial
epistemicide" in bioethics and consequently underscore some of its practical implications. I begin by emphasizing the weight of epistemic racism in bioethics by
referring to "whiteness" as a key concept of the current debate. I then introduce
the notion of epistemicide developed by de Sousa Santos and indicate potential
strategies for overcoming it. In particular, I adopt the approach of the "gatekeepers of bioethics" to analyze the factors through which the production of knowledge in bioethics is created, delimited and controlled. I conclude that the (re)ac-
tion of bioethics to racism must first be epistemic, as racism is deeply rooted in
our society and emanates from the formation of knowledge.

1 Introduction

Something[1] changed after the killing of George Floyd on May 25th, 2020, by a policeman associated with the Minnesota Police Department in the United States. It
is too early – and maybe too naïve – to say whether the wave of indignation
raised by this event will lead to definitive change. However, despite occurring
in the middle of the COVID-19 pandemic, that episode led to long-lasting pro-
tests. These, thanks to the resonance box of social networks, triggered a strong
reaction and global social condemnation based on the last words pronounced

---

1 An extended version of this essay has been published in Italian as: Carlo, Botrugno (2022):
"Dalla violenza razzista al razzismo epistemico: la bioetica si interroga". In: *Jura Gentium* 2.

Carlo Botrugno, University of Florence

https://doi.org/10.1515/9783110765120-004

by Floyd before dying: "I can't breathe". This sense of suffocation has indeed permeated a large part of our society, which took to the streets to support the Black Lives Matter (BLM) movement, an international social movement that originated in the United States in 2013 with the purpose of reacting to police brutality and racist violence against black people.

Seemingly, the same sense of suffocation has finally also permeated bioethics. Mere days after the death of George Floyd, bioethics scholars wondered about their effective commitment to fighting racism, i.e. to reacting to both implicit and explicit forms of persecution, discrimination, violence, prevarication, oppression, conditioning, bias, and any other expression of racism, which are usually perpetrated by white people to the detriment of black, indigenous, and other groups such as migrants, refugees, and the Roma people. In particular, some leading US and Canadian institutions in the field of bioethics and medical ethics took the initiative to emphasize the need for a deeper awareness of the degree to which prejudice, discrimination, and racism are rooted within our discipline. Several statements were adopted that expressed self-criticism, and multiple initiatives were launched with the aim of putting the notion of bioethics against racism "in action" (e.g. Association of Bioethics Programs Directors 2020; The Hastings Center 2020; Kennedy Institute on Ethics 2020; Nuffield Council of Bioethics 2020). In such a context, it has become evident that bioethics must wake up from a long sleep that (with few exceptions) has relegated our discipline to a powerful tool at the disposal of the "global north", i.e. rich countries, fluent societies, and well-educated and resourceful social groups and people.

Sadly, the killing of George Floyd is not an isolated case but rather the umpteenth example of a long history of violence against African-American citizens, in particular violence committed by police officers. However, racism is not exclusive to the United States but rather an urgent global matter. Another country in which racism too often leads to inconceivable violence and murder is Brazil. On November 20th, 2020, a Brazilian citizen named José Alberto Silveira Freitas was beaten to death by a military officer and a security guard in the parking lot of a supermarket. The 40-year-old brown man was brutally beaten by one assailant while the other kept him immobilized, meaning that Freitas suffered an impressive series of punches straight to his face. He died in a pool of blood outside the market itself. As in the case of George Floyd, the full episode was recorded on video. Compared with the former, however, this occurrence could seem even more striking, since none of the people witnessing that brutality intervened or protested, and they did absolutely nothing to help the victim, who was begging "please stop" during the beating.

These crude episodes clearly show that racism and, to some extent, even racist violence, is "naturalized" in many parts of our society. In other words, racism has a degree of "social legitimization", which means that for many people, the violence suffered by black, indigenous, brown, migrant, refugee, or Roma people is something less than "ordinary violence"; it is *minus quam* violence.

It is worth emphasizing the fact that racial profiling by policemen has already been recognized as a potential cause of adverse health outcomes. In this regard, Laurencin and Walker (2020) identified six different ways in which interaction with police can cause health damage. The first four of these ways are direct, while the other ones are indirect:

> (1) violent confrontation with police that causes injury or death; (2) police language that escalates a confrontation through micro-aggressions or macro-aggressions; (3) sub-lethal confrontations with police; (4) adverse health consequences of perceived or vicarious threat, i.e. the mere belief in potential harm by police injures health [...] (5) through knowledge of or personal relationship with someone who directly experienced racial profiling; (6) through public events without a personal knowledge of the unarmed person threatened or killed by police as a result of racial profiling, but where such events cause both individuals and the community at large to perceive a threat (Laurencin and Walker 2020, p. 393).

This classification shows that punches to the face or a policeman's knee on the neck are merely the most visible – and shocking – manifestations of racist violence. But racism can take several forms besides brutal violence and killing. Racism exists in more subtle and nuanced ways, which makes it hard to identify and oppose. Racism can be – and indeed is – veiled in politics, instilled in social and professional practices, and infused in our views and perspectives, and it sometimes occurs in a manner as naturally as wearing glasses. This is exactly where bioethics must intervene. Unveiling and fighting this hidden racism is a major goal of the medical, healthcare, and bioethics fields. In other words, bioethics must take action and contribute to reversing the direction of our racist societies, promoting equality and respect, thus defending everyone's right to exist. To achieve this purpose, however (as I argue herein), we must intervene at the root of racism in bioethics, i.e., the production of knowledge relevant to our discipline and related subject areas. It is indeed this epistemic ground that informs the emergence of policies, practices, views, and perspectives (both racist and anti-racist).

Based on the work of Boaventura de Sousa Santos (2007), in this chapter, I focus on the "racial epistemicide" of bioethics and consequently highlight some of its practical implications. In the second section, I emphasize the weight of epistemic racism by referring to a key concept in the current debate: the "whiteness" of bioethics. In the third section, I discuss the perilous ambivalence of the notion of race in biomedical research, particularly in genetics, alongside its (un)ethical implications. In the fourth section, I take up the approach of the "gatekeepers of bioethics" proposed by Chattopadhyay and colleagues (2013) and include additional factors by which the production of knowledge in bioethics is influenced. Through this analysis, I conclude that the (re)action of bioethics to racism must first be epistemic, as racism is deeply rooted in our society and emanates from the formation of knowledge, which must be always regarded as a "political act" (Andersen and Collins 2018).

2 How much does the epistemic racism in bioethics weigh?

The daily practice of medicine involves many episodes of racism (Tello 2017; Romano 2018), including prejudice, discrimination, oppression, and abuse based on people's identifiable visible markers. Arguably, only a very small number of these episodes are reported and shared, while the majority remain hidden, as invisible as the people who suffer from them. However, medical racism and discrimination can be considered to be merely the visible – and astonishing – face of the phenomenon. Metaphorically speaking, these episodes could be viewed as small leaves sprouting on the tree of racism. And where does this tree plunge its roots and get nourishment? My answer focuses on the production of knowledge or what can be called the epistemic ground on which bioethics arose and proliferated.

In his academic work, the philosopher and sociologist of law Boaventura de Sousa Santos emphasized the need to decolonize knowledge to stop epistemicide, i.e. the process by which "an immense wealth of cognitive experiences [other than the white-western] has been wasted" (Santos 2007, p. 37). Working on the connection between the production of knowledge and the historical mechanisms of dispossession – driven first by colonialism and then by capitalism – enabled Santos to emphasize the processes through which knowledge has been enclosed, delimited and controlled by the white elite to maintain privilege and supremacy acquired throughout history. As argued by Santos, "[g]lobal social injustice is therefore intimately linked to global cognitive injustice. The struggle for global social justice will, therefore, be a struggle for cognitive justice as well" (Santos 2007, p. 63). To overcome cognitive injustice, the author remarks on the need for epistemic efforts to shift from the "monoculture of modern science" to the "ecologies of knowledges" or to the "recognition of the plurality of heterogeneous knowledges (one of them being modern science) and on the sustained and dynamic interconnections between them without compromising their autonomy" (Santos 2007, p. 27). As Santos notes, the "ecology of knowledges" should be based on the idea that knowledge is "inter-knowledge" (Santos 2007, p. 27).

The work of Santos can be useful to understand how much cognitive injustice has permeated the production of knowledge in bioethics and related fields. A core element of the debate about racist bias in bioethics is currently represented by the notion of "whiteness". Such a qualification does not simply refer to "being white" as compared to "being black" or "non-white". Indeed, it has been noted that whiteness is "not necessarily about the skin color of those who practice bioethics. Rather, it is first and foremost about the dominant cultural norms and ideologies that have come to determine how bioethics is practiced, and what principles and contexts are considered relevant in the inquiry" (Ho 2016, p. 24).

Even more explicit is the argument made by Catherine Myser (2003), who views whiteness as "a marker of location or position within a social, and here racial, hierarchy – to which privilege and power attach and from which they are wielded – and how this is complicated by a forgetting of the history of whiteness in the United States and by its current invisibility" (Myser 2003, p. 2).

In such a perspective, whiteness is considered to be a hegemonic relation of power through which bioethical theories, knowledge, practices and principles are constructed, with the result of (re)producing white privileges and supremacy (Myser 2003, p. 3).

Whiteness is claimed to be so rooted in both the practice of medicine and bioethics that even its criticism could be seen as an expression of its dominance or, as argued by Myser herself, a mere "inoculation of difference":

> introducing sociocultural diversity and difference without recognizing, highlighting, and problematizing the dominance and normativity of whiteness in United States bioethics merely inoculates difference in bioethics, ignoring the question of against whose invisible and seemingly neutral norms such "difference" is defined (Myser 2003, p. 5).

Some scholars have reflected on the notion of whiteness and investigated the extent to which it shaped the identity of white healthcare professionals (Romano 2018; Gustafson 2007). These discerning self-portraits of whiteness in healthcare offer useful insights into the relationships that white professionals develop with and toward the "racialized other" (Gustafson 2007). For Gustafson, emphasis must be given to the epistemic implications of the dominance of whiteness, which is described as an "absent presence" in the construction of his own identity: "[m]y social location or, more precisely, my white identity influences what I see, the assumptions that focus my attention, the observations that I make, the problems I identify, the solutions that I generate and, more broadly, the knowledge that I produce" (Gustafson 2007, p. 155).

Conversely, from the perspective of black and other non-white people, in a field of bioethics that is dominated by whiteness, the former would be paradoxically subject to both "hypervisibility", given that "their medical trait is made a defining characteristic of their existence", as well as to "medical erasure", which refers to the circumstance that "their medical needs are left unaddressed and ignored" (Konnoth 2020, p. 1). On this perspective, bioethics would need to decenter "whiteness" or remove it as a predominant point of reference for the production of knowledge relevant to our discipline.

Racial-based epistemic inheritance is also a core element in the work of Katya Gibel Azoulay[2], who argues that "we have inherited a set of racial theories that articulated an anxious need to rationalize separation even as interaction and mixing were producing new people that defied and refuted the theories under construction" (Azoulay 2006, p. 371). However, while for some scholars, whiteness is "not only" a matter of skin color, on Azoulay's view, there is no room for an epistemic category that is based on – and thus acknowledges – the possibility of differentiation and identification of a particular skin color. From such a perspective, whiteness itself would be a product of the "western academy", particularly of bioethics in the U.S., that would be used as an epistemic scapegoat to avoid the contradictions and inconsistencies inherent in ensuring that skin color rather than class remains main reference for analysis concerning epistemic racism.

4 The discreet charm of the notion of race in genetics

As soon as vaccines for COVID-19 started to be tested on humans, concerns were raised regarding the call for "racial inclusivity" in their development (Campbell 2020). The recruitment of certain population groups that were labelled as "ethnic minorities" was indeed driven by the idea that COVID-19 vaccines "must work" among those populations as well. However, as argued by Colleen Campbell, such an idea is imbued with medical racism given that race does not have any genetic significance. This confirms that COVID-19 management and many (scientific) discourses emerging from the pandemic are rife with biological notions of race.

---

2 Until 2007, Katya Gibel Mevorach published under the name Katya Gibel Azoulay.

Without denying the need for a broadly effective vaccine against COVID-19, Campbell argues that "recruitment strategies that reinforce biological race risk justifying the disparities they are designed to address; they also harm people of color by reifying 'stigmatizing notions of racial difference'" (Campbell 2020, p. 1). This entails that certain population groups, such as African-Americans, are constantly "being framed as biologically at risk because of race" (Campbell 2020, p. 1)

Other scholars have highlighted the risk of referring to racial and ethnic categories as a means of taking differences into account. One such researcher is Patricia King, who emphasizes the fact that

> [i]nvestigators are now required to categorize study participants using self-identification into racial and ethnic categories used in the United States Census. Subjects must be offered the opportunity to select more than one racial designation. This well-intentioned effort to include African Americans in research, however, risks re-enforcing the now-discredited belief that "race" is a biological category and allowing biological or genetic differences to define racial and ethnic groups that are actually socially constructed. Using racial variables in research even for laudatory goals must be done carefully, otherwise the question will remain, "is this racial"? (King 2004, p. 152)

Racial inclusivity reminds us of the "inoculation of difference" in bioethics that was evoked by Catherine Myser (2003), i. e. a mechanism through which whiteness incorporates racialized factors with a view to exerting its control over them. As noted by Katya Gibel Azoulay in this context, "[a]lthough genetic studies are promoted as a new field attentive to genetic codes that invalidate race as biology, the concept of 'race' continues to be deployed as an organizer of difference" (Azoulay 2006, p. 359). The notion of race indeed maintains a privileged relationship with genetics, turning phenotypical markers into a principle of social and epistemic construction of the world. However, Azoulay could not be more explicit in rejecting such a perspective: "in the matter of race, there is no such thing as 'getting it right'. There are no generic races precisely because race is a metaphor, a social construct: a human invention whose criteria for differentiation are neither universal nor fixed but have always been used to manage difference" (Azoulay 2006, p. 358).

Therefore, despite the fact that the biological notion of race should already have been dismantled and relegated to a (racist) metaphor (Schwartz 2001) or a social construct (Karsjens and Johnson 2003), the relationship between race and genetics continues to exert its "discreet charm" on biomedical research, and consequently also on the practice of medicine and clinical ethics. Several scholars indeed defend the value of racially-based distinctions in clinical research. Among this group is Jay N. Cohn, who argues that this discussion should not focus on the existence of population differences, but rather on "how to describe those differences with more precision" as "distinctions does not make the differences disappear". From this perspective, he suggests "working towards better approaches in dealing with the differences, not raising legal and moral arguments" (Cohn 2006, p. 553). Accordingly, Anton van Niekerk (2016) proposes "to draw a distinction between 'racism', which is obviously contemptible and is outright rejected as a relevant category for biomedical research, and 'racialism' which might encapsulate a relevant criterion for legitimate group identification" (van Niekerk 2016, p. 269). van Niekerk thus evokes Ken Jones' definitions of racism and racialism, where the first alludes to "a belief or doctrine that inherent hierarchical differences among the various human races determine cultural or individual achievement, usually involving the idea that one's own race is superior" (Jones 2012, p. 1); by contrast, racialism would be defined as "a belief in the existence and significance of racial differences, but not necessarily that any hierarchy between the races exist" (Jones 2012, p. 1). On the basis of this distinction, Jones suggests considering race to be "a complex and multi-layered notion with multiple components" (Jones 2012). Therefore, future research should not "discriminate on the basis of race", but neither should it completely disregard race as a "factor that sometimes does seem to be relevant" (van Niekerk 2016, p. 271)

The usefulness of resorting to the notion of race and its categorizing power – alongside its "perilous ambivalence" – also appears as a concern in this book. In chapter 4, Celia Mariana Barboza de Sousa and colleagues show that taking into account "population aspects", although framed in terms of skin color, can be a vector to account for the health needs of black and brown populations in Brazil, i.e. in a socio-cultural context that has been historically marked by racist attempts of invisibilizing these groups. The authors indeed advocate that further knowledge drawn from studies concerning the black and brown population could enhance policy actions aimed at protecting health of black and brown populations.

From a different perspective, in chapter 5, Nchangwi Syntia Munung illustrates the (ethical) pitfalls of resorting to such categories as ancestry, ethnicity, and geographical location in genetic studies related to African populations. As she remarks, using ethnicity as a "proxy" for genetic research would require a precise definition of the ethnic groups in question. However, this could be highly problematic in Africa, where a same person could belong to several ethnicities, and these ethnicities are neither clearly defined nor even univocally associated with phenotypical features, instead often serving as mere tokens of a peculiar cultural heritage. In such a context, Munung problematizes the use of these labels as descriptors in population studies as they can feed the negative connotations that are sometimes intrinsic in their creation and spread. She indeed claims that the "social engagement" of researchers is necessary in order to balance the (scientific) need to use to these labels with their potential impact on society.

On different grounds stands the "argument" made by Katya Gibel Mevorach in chapter 6, who is not very concerned with the "political correctness" of racial- or ethnic-based descriptors but rather with their intrinsic significance. Mevorach could not be more explicit when affirms that "in the area of medical research, diagnosis and treatment, categorizations of human beings that infer racial taxonomy, racial categories and racial derivatives should be relegated to the historiography of scientific racism and viewed as 'bad science'" (Mevorach, p. 14). This entails that researchers must be committed to creating "new data schemes" (Mevorach, ibid.) for biomedical research.

To support Mevorach's argument, it is worth remembering that "race" is genetically insignificant not merely with respect to the identification of black or brown people, as she clearly illustrates. A study conducted a few years ago by Richard Tutton (2007) in the field of population genetics explicated notions of whiteness that are often embedded in scientific grounds related to the inclusion or exclusion of "ethnic groups" on the basis of genetic differences. As Tutton notes, "[a]ssumptions about the genetic homogeneity of whites/Caucasians significantly underpin the research design and rationale of [population genetics] studies" (Tutton 2007, p. 567). However, as he emphasizes, "such assumptions would seem to be problematic as in some instances evidence from genetic analysis could challenge the idea that whites/Caucasians are indeed all that homogeneous" (Tutton 2007, p. 567). In other words, population groups are currently being "geneticized" as having some degree of clinical relevance, while research pertaining to genetics discloses "new ways of categorising people that transcend traditional racial and ethnic classifications based on shared genotypes relevant to understanding disease risk rather than on crude proxies such as skin colour" (Tutton 2007, 567).

5 The epistemic gatekeepers of bioethics

Research conducted several years ago by Chattopadhyay and colleagues (2013) assessed the composition of the editorial boards of 14 leading scientific journals in the field of bioethics and medical ethics. The result highlighted the severe underrepresentation of scholars from developing countries. By categorizing these countries on the basis of their Human Development Index (HDI), the authors discovered that

approximately 95 percent of editorial board members are based in (very) high-HDI countries, less than 4 percent are from medium-HDI countries, and fewer than 1.5 percent are
from low-HDI countries. Eight out of 14 leading bioethics journals have no editorial
board members from a medium- or low-HDI country. Eleven bioethics journals have no
board members from low-HDI countries (Chattopadhyay et al. 2013, p. 7).

In light of this result, the authors argued that the editorial boards of these journals would act as "gatekeepers" of bioethics. This would explain why "bioethics
pays more attention to esoteric ethical problems facing wealthy nations than it
does to issues such as poverty, hunger, and health inequities that are global
in nature" (Chattopadhyay et al. 2013, p. 8).

I find the gatekeepers perspective particularly appropriate with respect to
analyzing the epistemic factors from which the production of knowledge in bioethics and related fields (as well as its racial epistemicide) stems. Without any
intention to be exhaustive, I think that medical education, particularly bioethics
courses, and ethical committees (ECs) and institutional review boards (IRBs) can
be included today among the epistemic gatekeepers of bioethics.

With respect to medical education, it is clear that bioethics courses and ethical training can play a major role in providing our healthcare professionals and
bioethicists with the competences required to unveil and react to implicit racist
biases in clinical practice and ethical consultancy. As argued by Danis and colleagues (2016), bioethics teachers are indeed

well suited to helping students appreciate that individuals each have a unique history, view
of the world, and life plan that is not merely a function of their race or culture. This more
complex understanding of the relationship of an individual to her or his culture, is particularly important to teach to students in the health professions who will be in a position to
pursue shared decision making with their patients (Danis et al. 2016, p. 7).

However, an initial problem in this regard is that, with the exception of Anglo-
European countries, the penetration of bioethics and medical ethics courses
globally is highly fragmented and locally contingent (Yung Ngan & Hiong Sim
2020). In the Asian context, for instance, it has been emphasized that "[u]niver-
sities have a limited capacity to provide professional training of ethics and moral
philosophy to sustain teaching quality" (Yung Ngan and Hiong Sim 2020, p. 38).
Moreover, a further challenge is related to the "limited curriculum time to include ethics in the packed medical teaching timetable" (Yung Ngan and Hiong
Sim 2020, p. 38).

The need to strengthen bioethics and medical ethics teaching in some regions of the world is also shown by the efforts of UNESCO, with Bioethics Chairs
activated in several countries in the global south, such as countries in Latin

America and the Caribbean (Hall and Tandon 2017; De Lopez et al. 2016). In this region, bioethics has become intertwined with the enhancement of human rights and a strengthening of democracy. The result is a "more politicized" concept of bioethics (Cunha and Lorenzo 2014; García Alarcón 2012) which is also expression of the need to protect local populations from the commercial interests of medical care firms. As has been widely acknowledged (Lorenzo and Garrafa 2011), some population groups must be protected from attempts to leverage poverty to stimulate their recruitment in medical trials.

Aside from the problem of incorporating bioethics and medical ethics within medical education, differences also exist in the degree of sensibility for the racial issues and the respect for diversity that available courses provide to their students (Greenberg et al. 2017). As seen above (section 4), there are significant differences among scholars concerning "how" to fight racism in medicine and healthcare and thus also in bioethics. Nevertheless, commitment to fighting racism remains locally contingent and therefore highly variable. In some cases, such a sensibility has been absent or insufficient even in top-ranked bioethics teaching institutions, as also proved by the aforementioned statements published in the US and Canada following the death of George Floyd.

On the basis of a 2019 report by the Association of American Medical Colleges, Klugman (2020) highlighted that in the U.S. medical schools, "only 3.6% of the faculty identify as black or African-American despite being 14% of the U.S. population" (Klugman 2020, p. 1). On this account, Klugman remarks that "[e]xpecting this small number of faculty to take on the duty of changing the culture of medicine is unrealistic and unfair" (Klugman 2020, p. 1).

All this confirms the need to profoundly rethink medical school curricula to make them more compliant and more committed to fighting racism and reducing health inequalities and discrimination. In this regard, the idea of a "medical school social justice curriculum" has been already launched (Coria et al. 2013) alongside medical training programs, with the aim of enabling "students to recognize and redress adverse medically relevant social factors" (Coria et al. 2013, p. 1448). Even such a proposal, however, cannot truly be successful until bioethics teaching is not "decolonized" from its hidden and intrinsic racist epistemology. To pursue this goal, I agree (once again) with Katya Gibel when she argues that our role as professors is to allow our students to "explicitly situate their knowledge and [be] attentive to their learning as a process" but also "to pay attention to why information and analysis may seem persuasive or irreconcilable when it contradicts their certitude" (Mevorach, p. 118). In other words, we can continue to resort to social categories but not "fetishize" them because "[p]hysical features matter but racial designations are insufficient and opaque" (Mevorach, ibid.).

The gatekeeping role played by ECs and IRBs is another key element to consider. In this regard, Anita Ho (2016) already stressed that the ability of clinical ethicists to address racist issues adequately must not be taken for granted because it depends mostly on their "working environment". As she notes, clinical ethicists have much less "freedom" than academic bioethicists. This makes it difficult for some of the former "to challenge the unequal health care system and discriminatory practices that continue to disadvantage certain ethnic groups, especially outside the scope of official consultation requests" (Ho 2016, p. 23).

To act immediately while continuing to draw attention to anti-racist education in bioethics and medical ethics, one idea could be the establishment of committees focused on ethnic and racial issues that contribute to the task of addressing race-related issues and promoting sensitivity regarding this matter, including with respect to public opinion. In the United States, there are several examples of these committees, such as the Ethnic and Racial Issues Committees established by the Society for Research in Child Development (SRCD), which is "responsible for the development and oversight of activities pertaining to the participation of minority scholars in the [SRCD] and for promoting developmental research on ethnic minority children and adolescents"[3]. Another example is the National Committee on Racial and Ethnic Diversity (NCORED) established by the National Association of Social Workers (NASW), whose aims are to "develop, promote and/or collaborate on methods of insuring inclusion of racial and diversity issues on NASW policies and programs [and to] promote the development of knowledge, theory and practice as related to racial/ethnic diversity". The Committee also monitors "policy changes, and data affecting policy changes, with regard to racial and ethnic groups both native-born and immigrant" and identifies "ways to eliminate racist and ethno-centric social work practices and policies and make recommendations to appropriate organizational units for action"[4].

6 Conclusions

The COVID-19 pandemic has confirmed that socio-economic disparities and racial discrimination act as determinants of health and therefore affect epidemiological distribution in contemporary societies dramatically (e. g. Botrugno 2020;

---

3 See https://www.srcd.org/about-us/who-we-are/committees/ethnic-racial-issues.
4 See https://www.socialworkers.org/About/Governance/National-Appointments/National-Committees/National-Committee-on-Racial-and-Ethnic-Diversity

Robinson et al. 2020; Greenaway et al. 2020). However, in spring 2020, it was already evident that black and Latino people were disproportionately affected by the impact of the virus across all age groups (Oppel et al. 2020; Greenaway et al. 2020). This difference is not explainable in virtue of genetic factors, thus representing another brick in the wall of the body of knowledge related to the weight of the social determinants of health and health inequalities (Botrugno 2018). As shown by a robust and transversal body of evidence, not only racism and discrimination but any processes of social stigmatization – particularly those related to the perpetration of hegemonic power relations – can push some groups to the margins of society, thereby provoking a socially-informed epidemiological distribution that takes the form of "structural violence" (Farmer 2004). From this perspective, it becomes clear that this is not solely an issue of epidemiology but rather a matter of global health justice, thus making it an urgent topic for bioethics. However, as argued herein, the awakening of bioethics must first be epistemic. There is a need for a cultural shift through which the dominance of approaches and perspectives that gave rise to discriminatory attitudes, behaviors and practices can be identified, challenged – or "historicized and contextualized", to use words of Katya Gibel Mevorach – and therefore rejected. We already know that the mere integration of differences and diversity into existing theoretical frameworks in bioethics and medical ethics is insufficient to eradicate racism from institutional policies and everyday healthcare practices. Whiteness itself, as a notion that supposedly serves to identify and fight racism, perpetrates a racist vision in which humans are categorized on the basis of their skin color. If such a notion seems to be useful in some contexts – e. g. with respect to discrimination against and the oppression of African-American or black and brown people in Latin America – it is of no significance for non-color based experiences of racism, such as the multiple phenomena of racism and discrimination found in countries in Central Europe regarding citizens from Eastern Europe or even in Eastern European countries, where citizens from neighboring countries are racialized, discriminated against and marginalized.

In light of this, there is a need to develop new epistemic notions and categories based on factors that cannot be traced back to the "historiography of racism" (see Mevorach in this book). This is fundamental to start to dismantle the hegemonic relationships and structures of power that enable the racial epistemicide of bioethics and thus facilitate the perpetration of discrimination, oppression, and social injustice.

References

Andersen, Margaret L.; Collins, Patricia H. (2018): *Race, Class, and Gender: An Anthology*. Cengage: Boston, Massachusetts.

Association of Bioethics Programs Directors. (2020): *Statement on Violence, COVID, and Structural Racism in American Society*. https://www.bioethicsdirectors.net/abpd-state ment-on-violence-covid-and-structural-racism-in-american-society/, visited on 03 August 2022.

Azoulay, Gibel Katya. (2006): "Reflections on race and the biologization of difference". In: *Patterns of Prejudice* 40. N. 4–5, pp. 353–379.

Botrugno, Carlo. (2018): "Healthcare, migrations and everyday bioethics: Weighing the difference". In: *L'altro Diritto* 1. Pp. 91–118.

Botrugno, Carlo. (2020): "El papel de la tecnología en la gestión de la pandemia de CoViD-19". In: *RedBioética Unesco* 21.Pp. 13–20.

Campbell, Colleen. (2020): "Racial inclusivity in COVID-19 vaccine trials". In *Bill of health*. https://blog.petrieflom.law.harvard.edu/2020/09/22/racial-inclusivity-covid19-vaccine-tri als/, visited on 03 August 2022.

Chattopadhyay, Subrata; Myser, Catherine; De Vries, Raymond. (2013): "Bioethics and Its Gatekeepers: Does Institutional Racism Exist in Leading Bioethics Journals?". In: *Bioethical Inquiry* 10. N. 1, pp. 7–9.

Cohn, Jay N. (2006): "The use of race and ethnicity in medicine: lessons from the African Heart Failure Trial". In: *J Law Med Ethics* 34. N. 3, pp. 552–554.

Coria, Alexandra; McKelvey, Gregory; Charlton, Paul; Woodworth, Michael; Lahey, Timothy. (2013): "The Design of a Medical School Social Justice Curriculum". In: *Acad Med* 88. N. 10, pp. 1442–1449.

Cunha, Thiago; Lorenzo, Cláudio. (2014): "*Bioética global na perspectva da bioétca crítica*". In: *Revista Bioética* 22. N. 1, pp. 116–125.

Danis, Marion; Wilson, Yolanda; White, Amina. (2016): "Bioethicists can and should contribute to addressing racism". In: *AJOB* 16. N. 4, pp. 3–12.

De Lopez, Verges Claude; Sánchez, Delia; Garrafa, Volnei; Peralta-Corneille, Andrea. (2016): "The Impact of the UNESCO International Bioethics Committee on Latin America: Respect for Cultural Diversity and Pluralism". In: Bagheri, A., Moreno, J., Semplici, S. (eds) *Global Bioethics: The Impact of the UNESCO International Bioethics Committee. Advancing Global Bioethics*, vol 5. Cham: Springer.

Farmer, Paul. (2004): "An anthropology of structural violence". In: *Current Anthropology* 45. N. 3, pp. 305–325.

García Alarcón, Rodrigo Hernán. (2012): "A bioética na perspectiva latino-americana, sua relação com os direitos humanos e a formação da consciência social de futuros profissionais". In: *Revista Latinoamericana de Bioética* 12. N. 2, pp. 44–51.

Greenaway, Christina; Hargreaves, Sally; Barkati, Sapha; Coyle, Christina M., Gobbi, Federico; Veizis, Apostolos; Douglas, Paul. (2020): "COVID-19: Exposing and addressing health disparities among ethnic minorities and migrants". In: *J Travel Med* 27. N. 7, taaa113. doi: 10.1093/jtm/taaa113.

Greenberg, R. A., Kim, C., Stolte, H., Hellmann, J. Shaul, R. Z., Valani, R., & Scolnik, D. 2016. Developing a bioethics curriculum for medical students from divergent geo-political regions. *BMC Med Educ* 16, 193.

Gustafson, Diana L. (2007): "White on whiteness: becoming radicalized about race". In: *Nursing Inquiry* 14. N. 2, pp. 153–161.

Hall, Budd L.; Tandon, Rajesh. (2017): "Decolonization of knowledge, epistemicide, participatory research and higher education". In: *Research for All* 1. N. 1, pp. 6–19;

Ho, Anita. (2016): "Racism and Bioethics: Are We Part of the Problem?". In: *AJOB*. 16. N. 4, pp. 23–25.

Jones, Ken. (2012): "Racism vs Racialism". In: *The Real Ken Jones*. https://therealkenjones. wordpress.com/2012/03/19/racism-vs-racialism/, visited on 03 August 2022.

Karsjens, Kari L.; Johnson, Joanna M. (2003): "White normativity and subsequent critical race deconstruction in bioethics". In: *AJOB* 3. N. 2, pp. 22–23.

King, Patricia A. (2004): "Reflections on Race and Bioethics in the United States". In: *Health Matrix* 14. N. 1, pp. 149–153.

Klugman, Craig. (2020): "Our Response to Racism Should Not Be More Unpaid Work for Black Faculty". In: *Bioethics Today*. http://www.bioethics.net/2020/06/our-response-to-racism-should-not-be-more-unpaid-work-for-black-faculty-part-i/, visited on 03 August 2022.

Konnoth, Craig. (2020): "The Double Bind of Medicine for Racial Minorities". In: *Bill of Health*. https://blog.petrieflom.law.harvard.edu/2020/10/13/double-bind-medicine-racial-minorities/, visited on 03 August 2022.

Laurencin, Cato T.; Walker, Joanne M. (2020): "Racial Profiling Is a Public Health and Health Disparities Issue". In: *Journal of Racial and Ethnic Health Disparities* 7. N. 3, pp. 393–397.

Lorenzo, Cláudio; Garrafa, Volnei. (2011): "Ensayos clínicos, Estado y sociedad: ¿dónde termina la ciencia y empieza el negocio?". In: *Salud Colectiva* 7. N. 2, pp. 166–170.

Marmot, Michael. (2015): *The Health Gap. The challenge of an unequal world*. London: Bloomsbury.

Myser, Catherine. (2003): "Differences from Somewhere: The Normativity of Whiteness in Bioethics in the United States". In: *AJOB* 3. N. 2, pp. 1–11.

Niekerk Anton. A. van. (2016): "Non-racism as a core value in bioethics". In: *Ethics, Medicine and Public Health* 2. N. 2, pp. 263–271.

Nuffield Council of Bioethics. (2020): *A perilous moment for our nation*. https://www.thehas tingscenter.org/news/a-perilous-moment-for-our-nation/, visited on 03 August 2022.

Oppel, Richard A. Jr.; Gebeloff, Robert; Lai, Rebecca K. K.; Wright, Will;. Smith, Mitch. (2020): "The Fullest Look Yet at the Racial Inequity of Coronavirus". In: *New York Times*, 05 July 2020. https://www.nytimes.com/interactive/2020/07/05/us/coronavirus-latinos-african-americans-cdc-data.html, visited on 13 July 2022.

Robinson, Laura; Schulz, Jeremy; Khilnani, Aneka; et al. (2020): "Digital inequalities in time of pandemic: COVID-19 exposure risk profiles and new forms of vulnerability". In: *First Monday* 25. N. 7, ff10.5210/fm.v25i7.10845 ff.

Romano, Max J. (2018): "White Privilege in a White Coat: How Racism Shaped my Medical Education". In: *Annals of Family Medicine* 16. N. 3, pp. 261–263.

Schwartz, R. S. 2001. Racial profiling in medical research. *N Engl J Med.* 344, pp. 1392–1393.

Santos, Sousa Boaventura de. (2007): "Beyond Abyssal Thinking: From Global Lines to Ecologies of Knowledges". In: *Review* 30. N. 1, pp. 45–89.

Tello, Monique. (2017): "Racism and discrimination in healthcare: Providers and patients". https://www.health.harvard.edu/blog/racism-discrimination-health-care-providers-pa tients

The Hasting Center. (2020): *A Perilous Moment for Our Nation*. https://www.thehastings
  center.org/news/a-perilous-moment-for-our-nation/, visited on 13 July 2022.
Tutton, Richard. (2007): "Opening the white box: Exploring the study of whiteness in
  contemporary genetics research". In: *Ethnic and Racial Studies* 30. N. 4, pp. 557–569.
Yung Ngan, Olivia Miu; Hiong Sim, Joong. (2020): "Evolution of bioethics education in the
  medical programme: a tale of two medical schools. In: *International Journal of Ethics
  Education* 6. Pp. 37–50. https://doi.org/10.1007/s40889-020-00112-0.

Robin Pierce
Interrupting Technological Pathways to Health Inequities

Abstract: Technological innovation in healthcare wields considerable influence on patient care and health outcomes. From the introduction of dialysis to the recent push of artificial intelligence (AI) into the healthcare domain, technology is widely seen to hold the secret to improved outcomes and increased longevity. While the benefits of tech innovation have resulted in improved outcomes of multiple sorts, the degree to which technology confers benefits is not experienced equally across groups. Health disparities along racial lines have been documented for decades, and despite relentless evidence of health disparities, these inequities persist. Given the reliance on technology in healthcare, close examination of the role of technology in the creation and perpetuation of health inequities in the provision of healthcare is essential. The urgency of holding technological innovation under scrutiny is perhaps even more pronounced in the digital age when so much of the clinical experience is becoming digitized. This expanded digitization in healthcare also gives rise to new and different ways of creating and contributing to health inequities. As healthcare adopts ever more technological resources in the effort to improve health outcomes, the need for systematic scrutiny and accountability specifically directed at pathways to health inequities also intensifies. Joining the call for increased research on interventions, this chapter briefly examines the role of technological innovation in the creation and perpetuation of health inequities, and highlights the potential usefulness of regulation theory particularly pertaining to technology in developing effective strategies to interrupt technological pathways to health inequities as an essential complement to addressing root causes of health inequities.

1 Introduction

Technological innovation in healthcare wields considerable influence on patient care and health outcomes. From the introduction of dialysis to the recent push of artificial intelligence (AI) into the healthcare domain, technology is widely seen to hold the secret to improved outcomes and increased longevity. While the ben-

Robin Pierce, University of Exeter

https://doi.org/10.1515/9783110765120-005

efits of tech innovation have resulted in improved outcomes of multiple sorts, the degree to which technology confers benefits is not experienced equally across groups. Health disparities along racial lines have been documented for decades, and despite relentless evidence of differences in health outcomes, these inequities persist. For all the advantage-conferring capacities, once deployed, technology too often serves as one of the pathways to health inequities. Given the pervasive and increased reliance on technology in healthcare, close examination of the role of technology in the creation and perpetuation of health inequities in the provision of healthcare is essential. The urgency of holding technological innovation under scrutiny is perhaps even more pronounced when so much of the clinical experience is becoming digitized. This expanded digitization in healthcare also gives rise to new and different ways of creating and contributing to health inequities. As healthcare adopts ever more technological resources in the effort to improve health outcomes, the need for systematic scrutiny and accountability specifically directed at pathways to health inequities also intensifies. In 2006, Cutler, Deaton, and Lleras-Muney predicted that there will be an acceleration in the production of new technologies in the coming years (Korda et al. 2011). However, several studies have also shown an increase in health inequalities upon the introduction of technological innovation (Weiss et al. 2018; Chang and Lauderdale 2009). Consequently, as both public health and clinical care increasingly adopt technological resources in the effort to improve health outcomes, the need for scrutiny also intensifies as to the potential and actual impact on health inequalities. The forms that this scrutiny can take are numerous, and have primarily included ethical, legal, and social issues (ELSI) analysis, various types of impact assessments, among others. However, approaches to interventions, and particularly to interventions and issues pertaining to health inequities, are inconsistent (see e. g. Char et al. 2020). A more systematic inquiry and policy approach for identifying and intervening in technological pathways to health inequities is urgently needed. This chapter explores technological pathways to health inequities, emphasizing a translational approach, complementary to ELSI, as a way of systematizing scrutiny and informing possible interventions, and highlights a more pronounced role for regulation theory in the development of appropriate interventions. The chapter begins with a brief description of structural racism and how it impacts health outcomes. I then situate technological innovation in the creation, perpetuation, and exacerbation of health inequities, citing examples of beneficial technological innovations that contributed to or exacerbated health disparities. The chapter then discusses technological pathways to health inequities as presenting possible intervention points, and concludes with brief consideration of how concepts from regulation theory could be and are strategically deployed to support systematic interruption of pathways to health inequities.

2 Racism, bias, and structural inequities

At the root of health disparities sits a complex web of causes, effects, and pathways that reflect and interact with multiple social systems including housing, labor, economic, credit, and education that cascade into a chain of cause and effect that results in unjust differences in outcomes for different racial and ethnic groups (Williams et al. 2019). When viewed globally, it becomes clear that the underlying causes (structural inequalities) operate contextually, but seem to follow a pattern of continuous and perpetual disadvantage for historically marginalized persons. Some causes are clearly related to structural and cultural racism in a country, while in other countries, the concept of race does little or no work in accounting for conscious or unconscious differential impacts. Often in its place are other categories of difference, such as ethnic origin, immigrant background, and socio-economic status (SES) that explain the disparate outcomes that mirror the manner of social "stratification" in a society. For example, in the US, race has historically occupied an important role in all facets of personal and professional life (see e. g. Gates 2020; Wilkerson 2020). While in Canada, the ethnic category of "First Nations", a term used to refer to indigenous populations, has played a prominent role in categorical marginalization and disadvantage (Pirisi 2015; Lafontaine 2018). Brazil presents a further example, showing the complexity and extent of racial marginalization and systematic disadvantage, as is discussed in Chapters 7 and 8 in this text. In Europe, social categories that are associated with disadvantage and marginalization vary from country to country but tend to track SES, ethnic origin, and immigration background (see e. g. Weiss et al. 2018). For example, in some countries immigrant populations, despite generations of presence in the country, still experience systematic disadvantage as a result (see e. g. Malmusi 2015).

While there are many different dimensions along which health inequities occur, this chapter will principally use the example of race to illustrate multilayered impacts and how different technological pathways manifest. Racism can be defined as

> an organized social system in which the dominant racial group, based on an ideology of inferiority, categorizes and ranks people into social groups called "races", and uses its power to devalue, disempower, and differentially allocate valued societal resources and opportunities to those defined as inferior (Williams et al. 2019, p. 2)

Thus, racism should be understood as a "system" that is enacted in multiple ways, operating through multiple pathways and reinforced by intersecting social systems throughout society. Three dimensions through which racism operates to disadvantage particular groups of people are 1) "structural racism", a term used to refer to "processes of racism embedded in laws, policies, societal practices that provide advantages to racial groups deemed superior", while, at the same time, oppressing, disadvantaging, or neglecting racial groups who are viewed as inferior, 2) "cultural racism", which refers to the embedded ideology of inferiority in language, values, symbols, imagery, and the assumptions of a society. Cultural racism is a pervasive, constant that results in a sort of ideological environment in which a view of the supposed inferiority of particular groups is embedded in a society, resulting in stereotyping, norms, implicit and explicit bias that reflect this view, and 3) "interpersonal racism", a term intended to capture bias that takes place in interactions among individuals (Williams et al. 2019) as, for example, a racial bias that implicitly or explicitly informs a physician's decision-making, behavior, or clinical assessments.

Much work has been done to explain the differences in health outcomes along racial lines (in countries where that characteristic is prominent), nevertheless, identification of "racism" as a causative force does not necessarily simplify our real-time understanding of the phenomenon as it impacts health profiles nor what, precisely, should be done about it. It does, however, diminish unsubstantiated claims about innate biological differences among races as the causative factor in poorer health outcomes (see e.g. Beckwith et al. 2017; Williams et al. 2019) and, perhaps most importantly, it points in the direction of structural and cultural inequities as a target for interventions. In recognition of various features related to the complexity of racism as a causative factor, Williams and colleagues speak in terms of the multi-layered and multi-dimensional nature of racism and its impact on health outcomes. They refer to a "foundational" layer in which critical causal factors generate an outcome, e.g. racism as a fundamental cause of racial inequities in health, which can be contrasted with "surface" layers that are associated with the outcome, but changes here do not result in changes to the outcome (Williams 1997). Within this framework, technology is often neither, but instead may represent an intermediate layer between foundational and surface layers and, moreover, constitutes a route from the foundational layer to a surface layer – a pathway.

While the temptation to regard technology as neutral has diminished, the reality is that technological innovation, however well-intended and beneficial in nature, does not affect health outcomes equally (e.g. Braun 2014; Benjamin 2019). That is, when a new technology is introduced into health care that has been shown to improve outcomes in both research and health care delivery, the improvement in outcomes is not necessarily uniform across ethnicities or races. This leads to the question of what is it about technological innovation that operates as a pathway to health inequities? Where in the process of pre-clinical and clinical translation, integration, and use, does racism gain traction leading to disparate consequences for patients? While increasing attention is rightly being devoted to dismantling the foundational causes of structural and cultural racism, technological pathways arguably operate as a conduit for structural and cultural racism in multiple ways. This chapter advocates for complementary strategies that address or interrupt pathways "while" efforts to dismantle racism continue and increase. Indeed, as this chapter aims to show, technology often operates as a minion of structural racism becoming an increasingly influential piece of social systems through which structural racism wreaks its harm on historically marginalized and disadvantaged groups. Moreover, without scrutiny, technology can literally become a component of structural racism, a technological artifact that, in purportedly neutral fashion, perpetuates the social hierarchy in which it is embedded. Given the growing pervasiveness of technology in health care, it presents as a critical pathway for health inequities and, as such, warrants systematic and dedicated attention.

The magnitude and expansiveness of foundational causative factors necessitate complementary approaches that may serve to reduce inequities as part of a coordinated effort. As health disparities scholar, Nancy Krieger has observed, "structural problems require structural solutions" (Krieger 2021, p. 301). But until and while those structural problems are effectively addressed, of which technology may operate as one, the spectrum of pathways arguably present as points of intervention, as in employment, housing, and other social systems, have become sites of "pathway intervention", out of a recognition that the eradication of problematic attitudes and behaviors, resulting in structural and cultural racism is a mammoth undertaking and that we need not suffer preventable casualties until this goal is achieved.

3 Technological innovation and health inequities

In 2008, Glied and Lleras-Muney showed that "more educated individuals have a greater survival advantage in those diseases for which there has been more health-related technological progress" (Glied and Lleras-Muney 2008, p. 748). In other words, technological development is of greatest benefit to those who are already advantaged, with less benefit going to less educated persons. The COVID-19 crisis clearly illustrated the domino effect of structural racism in action (Yancy 2020). Allocation of life-saving technology has been often determined by life-expectancy once the technologically-supported treatment has been administered. However, structural racism leads to lower life expectancy through multiple pathways, therefore, resulting in profound inequities in the use of life-saving treatment and resulting health outcomes through a technological pathway (White and Lo 2021). A failure to recognize the structural racism inherent in the criteria leads to the perpetuation of disadvantage and inequity. If a technological innovation is used in healthcare based on indication, need, and suitability for treatment of presenting individuals, then there should be no discernible difference in outcomes that fall along lines other than the indicative criteria (when baseline health status and relevant biological factors are comparable). Yet, multiple studies have shown that this simply is not the case, and that the differences in outcomes resulting from technological innovation can often be traced to structural inequalities that are persistent and pervasive in society. There is a growing body of evidence as to why this is happening, putting forth such explanations as use, access, social determinants of health operating to uneven affect, and reproduced social bias and inequities, among others (see e.g. Glied and Lleras-Muney 2008; Williams 2012; Williams and Wyatt 2015; Williams and Collins 2001; Weiss et al. 2018). In 2005, Goldman and Lakdwalla found that

> improvements in the productivity of health care disproportionately benefit the heaviest health care users. Since richer patients tend to use the most health care, this suggests that new technologies – by making more diseases treatable, reducing the price of health care, or improving health care productivity – could widen socioeconomic disparities in health (Goldman and Lakdwalla 2005, p. 24)

Essentially, they posit that those who use the most health care, reap the greatest benefits from technological innovation. This seems logical. Indeed, other scholars have observed that for those conditions where the new interventions are effective, technological progress may result in rising health inequalities (Chang and Lauderdale 2009; Glied and Lleras-Muney 2008). Asking what structural factors play a role in one's ability to increase one's resources point to a path of intersecting social systems of education, labour, credit, etc. that find racism as a root cause (Williams et al. 2019). Yet, differential use and access are only one set of pathways that is exacerbated by technological innovation. Technology, in some ways, can and has become an instrument of structural and cultural racism, a means that contributes to and facilitates continued racial inequities in health.

Medical technology poses particular challenges with regard to rising relative inequalities in health (see Korda et al. 2011). Weiss and colleagues observe that "as quantification of health (...) intensifies and innovative technologies become the cornerstone of this transition, the connection between technology and health

is garnering increased attention" (Weiss et al. 2018, p. 25) Evidence for technological pathways to health disparities has been documented by many health policy scholars. Cancer survival, for example, shows significant disparities that
widen for most racial and ethnic populations as cancers become more amenable
to medical interventions (Tehranifar et al. 2009).

In Europe the pattern is similar in that health disparities or inequities align
with social inequalities already present in society. Weiss and colleagues conducted an extensive review of the literature on innovative technologies and social inequalities in health that looked at studies on health inequalities and technological innovation from 1996 to 2016 (Weiss et al. 2018). They found that the studies
"demonstrate that individuals of higher SES are the first to adopt, and benefit
most from, the introduction of innovative technologies in health, creating social
inequalities in health where they were once very low or nonexistent" (Weiss et
al. 2018, p. 2). Furthermore, they found that in some cases the introduction of
technology even resulted in "inverted" inequalities (i.e. improved health outcomes moved from lower SES groups to higher SES groups) (Weiss et al. 2018).
Although what influences SES can vary from country to country, there are unmis-
takable patterns. Marginalized and ethnic and racial minority populations generally fare less well upon the introduction of innovative technologies in healthcare (Weiss 2018). In countries where race is not a relevant category, SES has
been shown to correlate with a range of disparities in health, as does ethnicity
and immigrant background (see e.g. Malmusi 2015; Weiss 2018).

Health disparities in technologically-mediated treatment and care of Alzheimer's disease also show that disparities along ethnic and racial categories
is not limited to a single country or region. In the U.S., non-hispanic black persons had roughly double the risk of underdiagnosis as non-Hispanic white persons (Gianattasio 2019). In Denmark researchers identified access as a factor in
disparities, noting that "a worrisome difference in access to anti-dementia treatment and care for dementia patients with an immigrant background, despite
similar levels of adherence compared with the Danish-born population" (Stevnsborg 2016, p. 509).

Genetics and genomics offer different types of pathways to inequities. Genetic technologies join other technological innovations in confronting challenges to
equal access, particularly as used for diagnostic purposes and risk identification
(see Smith et al. 2016). However, genetic technologies present additional pathways to health inequities that may operate outside the health context, impacting
social determinants of health. Purported genetic links to violence, intelligence,
and educability find their way into the public discourse affecting attitudes, beliefs, practices, and even policies (see e.g. Beckwith and Pierce 2018; Beckwith
et al. 2017; Duster 2004; Morris et al. 2013).

4 Pathways and mechanisms

Research in the field of health inequities has made significant strides in understanding what is behind the unjust differences in health, and specifically, how technology fits into the dynamic of perpetuating or facilitating health disparities, identifying a range of "sites" where causative factors stemming from structural racism directly or indirectly lead to health inequities. For example, studies have shown differences in multiple aspects of the tech-supported clinical encounter, including diagnostic rates and practices (Geiger 2003), treatment decisions (Vo et al. 2021), pain relief and medication (Anderson et al. 2009; Kennel et al. 2019), which suggest pathways involving resource allocation, referral practices, setting of indicative criteria, and human bias or ignorance in the use of the technology. Additionally, recent discussions regarding the use of race or other categorical criteria (see e.g. Flanigan et al. 2021; Krieger 2021) highlight the impact of socially constructed terminology, terminology that then must interact with technological innovation, e.g. digital technologies that rely on historical training data using racial designations.

Particularly when considering the role of technology in the dynamic leading to health inequities, it may be useful to think in terms of direct and indirect pathways. An example of a direct pathway from racism to health inequities might be differentially prescribing adequate pain medication because of a belief that black people do not feel pain as strongly as white people. An indirect pathway, in contrast, might include a practice that operates neutrally on its face, but does so against a backdrop of structural inequality that serves to perpetuate disadvantage among the already disadvantaged. An example of this can be found in AI-driven e-alerts for acute kidney disease, which are trained on historical data in which structural racism is embedded (Pierce et al. 2021). Technological pathways often operate indirectly, but also include instances of direct impact on health inequities as, for example, diminished ability to detect disease in non-white patients.

Indirect pathways abound in the production of health inequities. Poorer health states that would seem to follow from behavioral or lifestyle practices, may actually be rooted in some structural phenomenon operating almost invisibly but generating impact on seemingly neutral practices. The finding of a link between structural racism leading to residential segregation (Williams and Collins 2001), which has been shown to have a significant impact on health outcomes and can be technologically-mediated, illustrates the downstream effects of structural racism and the virtually infinite pathways that structural racism and its sequelae negatively and differentially affect people's lives (see Shonkoff et al. 2021; Cuevas et al. 2020).

Ignoring these background inequities on the basis of asserted neutrality has repeatedly been shown to result in compounding those inequities within the context of whatever innovation or practice is introduced. Legal scholar Rachel Moran analyzes the phenomenon of how the switch to remote learning during the COVID-19 pandemic intensified patterns of segregation and isolation for students in disadvantaged groups (e. g. race, ethnicity, poverty). Moran showed that "the burdens of shifting to online learning did not fall equally on all students" (Moran 2021, p. 605), that disadvantaged children faced the most obstacles to remote learning, and their schools had fewer resources to respond to abrupt school closures resulting in reduced curricula and less engagement with teachers (Moran 2021). This seemingly "neutral" beneficial innovation that allowed for the continuation of education for young people in pandemic, actually served to perpetuate and compound existing inequities, effectively serving as a pathway to inequity, including health inequity by virtue of contributing to known inequalities in social determinants of health that fuel health disparities.

In the latter instance, technology-intensive online learning was introduced with the clear goal and ability to facilitate learning for students who could not attend school in person due to the pandemic. Moran compellingly observes the differential impact of introduction of this beneficial technology, which is attributable, not to some discriminatory use of the technology, but to existing structural inequities that resulted in more obstacles to remote learning and fewer resources with which to respond to the crisis, leaving already disadvantaged students with reduced curricula and less teacher interaction (Moran 2021). The phenomenon of the introduction of seemingly neutral technologies having disparate impact along racial, ethnic, and SES groups, particularly in the health care domain, leading to poor health outcomes for racial minorities and economically disadvantaged persons, is one that haunts technological innovation (Weiss et al. 2018).

The complex dynamic in which the causative factors directly and indirectly lead to health inequities, by way of technological pathways joins the mounting evidence pointing to the need to identify and develop effective strategies for intervening to eliminate health disparities and minimize disparate impacts on health with regard to both causative factors and their pathways. In the following section, I set forth some considerations for developing strategies as a way of expanding the conceptual toolbox for addressing technological and technologically-supported pathways to health inequities.

5 Technological pathways

Some have observed that social inequalities in health seem to be on the rise in high income countries (HIC) during a period of increasing technological innovation (Mackenbush 2012; Piot 2012; Weiss et al. 2018). Recent efforts to move in the direction of personalized medicine, along with reliance on AI-driven clinical decision support systems and the pervasive influx of digital technologies in both clinical and public health raises legitimate concerns about a resulting increase in social and racial inequities delivered through technological pathways. This echoes earlier concerns about the role of medical technologies in the creation and perpetuation of health disparities (Glied and Lleras-Muney 2008); Korda et al. 2011). Some scholars have referred to technological pathways in terms of a "re-production" of social inequalities in health (e. g. Lutley and Freese 2005). This certainly describes the end result in that health inequities mirror patterns of existing social inequities. However, there are arguably other mechanisms in action in the case of technological innovation. As Weiss and colleagues note, a broad foundation from which to further investigate and explain the connection between technological innovation and social inequities in health is still largely missing from the literature (Weiss et al. 2018).

AI presents a recent and compelling case of concern regarding the perpetuation and reproduction of structural inequities (see e. g. Vyas et al. 2020; Obermeyer et al. 2019; Eubanks 2018). Several scholars have pointed out the ways in which AI-driven technologies not only perpetuate structural inequities but also how it is positioned to create new categories subject to marginalization by the uncovering of correlations using big data that point to heretofore unknown combinations of characteristics associated with a particular trait or phenotype (see e. g. Wachter 2020), which can result in powerful potential pathways to health inequities being created with the uptake and integration of AI-driven technologies in health care. Even such seemingly benign AI-driven technologies as e-alerts for acute kidney injury (AKI) inadvertently carry forward structural inequities in the alert itself because of the metrics (how the baseline is formulated) built into the algorithm that is carried over from the KDIGO criteria (Pierce et al. 2021). Once taken up on a larger scale, this potentially beneficial medical advance becomes part of the infrastructure of structural racism and inequity, obscuring both the pathway to inequity as well as the opportunity to intervene. In short, technological innovation offers a plethora of benefits, but if the ongoing and escalating efforts to reduce and eliminate health disparities are to engage in optimally effective and optimally timed interventions, technological pathways must be given a place on that agenda.

Technology may operate as causative, perpetuating, or facilitating of health inequities through its design, operation, and use, yielding novel ways of impacting health care. For example, detection technologies that are trained primarily on white patients may show lower efficacy in detecting the same pathology in patients of color (Parika et al. 2019). In this way, deficiency in detection capabilities fall along racial and ethnic lines, thus "causing" unfair differences in treatment and, presumably, outcomes. Such deficiencies would logically be a human-generated derivative of structural racism. Whether the specific pathway is based on uneven use or availability of training data or failure to accommodate the obvious difference in phenotype among persons of different races, structural and cultural racism sits at the root. Causation could be a matter of design and, arguably, sits somewhere between foundational and surface causes (see Williams et al. 2019, where surface causes are distinguished from foundational causes in that if altered, surface causes do not change the impact on health inequities). Here, the technological feature that fails to function equally well in patients of different races is subject to change and, if removed or fixed, would result in a change in the impact of "this" technology on health inequities.

Like medical research that was historically primarily carried out on persons who were white and male (King 2002; Epstein 2008) the lack of diversity in clinical studies had downstream implications for health outcomes of those who were not white and male. Over time, involving more diverse study populations was prioritized and women and persons of color were actively recruited to trials, albeit with varying degrees of success.

5.1 Strategies to address technological pathways

It should again be emphasized that addressing technological pathways to health inequities does not in any way suggest that, as an interventional strategy, diminish the urgency of addressing root causes. Instead, addressing technological pathways should be viewed as among many possible complementary approaches to containing the impact of structural and cultural racism and its sequelae. Moreover, identifying the various pathways to health inequities is critical to creating effective interventions, and arguably to building in accountability in a timely manner. The combination of structural and cultural racism creates a pervasively permeating backdrop such that virtually any healthcare innovation will be seriously challenged in efforts to operate neutrally once adopted into health care delivery and practice. Thus, until the foundational causes of structural and cultural racism are effectively curtailed, unaddressed pathways will allow for and even facilitate the continuation of disparate impacts of technological innovation resulting in health inequities. As health disparities scholars have noted, there are virtually innumerable pathways to health inequities (Williams et al. 2019), arguably reflecting the range of points of contact. Wherever there is contact or interface between patients or the public and any aspect of the health-care system, a potential pathway can be found. Insurance, referrals, prescriptions, surgical procedures, counseling, emergency care, and so forth are all susceptible to operating as pathways to health inequity. These interaction points in healthcare that are facilitated by technology make technology a particularly pressing target, especially in view of pervasive and increased technological reliance in health care. This is potentially of profound consequence in the effort to eradicate health inequities, in that these inequities can and do result in unnecessary death and poor quality of life.

There are many ways to approach the issue of technological pathways to health disparities. The question is arguably no longer whether technology contributes to, perpetuates, or exacerbates inequities, but rather how technology does this, and how can these pathways successfully be interrupted. Questions regarding technology as a possible "instrument" of structural and cultural racism must be asked. It is not necessary that the technology be "weaponized" against a particular group of persons, intentionally or deliberately deployed in a discriminatory fashion. It is enough that it merely sits in or emerges within a pre-existing dynamic of structural privilege and disadvantage.

A causative factor such as racism, in general, is a difficult target in which to intervene effectively due to various reasons (see Yancy 2020), ranging from difficulties in identifying the problematic action or behavior, as might be the case in physician behavior or affect, to legal protections and prohibitions, as, in the case of changing attitudes when one's beliefs and views are protected by freedom of speech (thought). What can and should be prohibited is the expression of discriminatory beliefs and views in practice, as is addressed in anti-discrimination laws, e. g. employment and housing law (see McGowan et al. 2016 for discussion of law as a tool to reduce health inequities). Thus, while we can train, educate, and incentivize fair and non-racist behavior and practice in health care delivery, changing attitudes remains a difficult target.

Consequently, even though we may realize that structural and cultural racism is a foundational cause, and as a highly problematic and influential factor in producing health inequities, an approach that acknowledges and seeks to intervene in the numerous pathways by which racism inflicts its disparate impact seems well-advised. Recognizing this, health disparities scholar, Nancy Krieger (2021) has set forth some proposals for dismantling specific features of structural racism, e. g. minimal use of zip code and selective and justified use of race as a category. Such proposed interventions are designed to counter pathways (here, in research) that comprise structural racism and would be expected to have downstream effects.

6 Interrupting technological pathways – a pipeline view

Among the key issues regarding the development of effective strategies for addressing health inequities and their root and contributory causes is identifying the optimal timing for intervention. Recognition of this has resulted in a lively ELSI agenda frequently accompanying technological innovation, giving rise to the widely adopted view that ethical, legal, and social consideration should precede and accompany development and introduction of technological innovation. The Human Genome Project empaneled an ELSI committee at the outset of the mapping of the human genome (Collins et al. 1998). Other technological innovation has followed suit and, recently, has resulted in a (justifiable) burgeoning ethical and legal scrutiny of AI in health care, for example. Notably, this scrutiny is rightly directed not only at the adoption, integration, and use of AI, but also at the design (and training) of the technology itself, as well (see e.g. Char et al. 2020; Katell et al. 2020; Koenecke et al. 2020).

Nevertheless, one of the shortcomings of ELSI as an approach to addressing health inequities is that, while often incisive, it is very much ad hoc and relatively unsystematic. Responsible Research and Innovation (RRI) (Owen et al. 2012), a parallel approach taken up in Europe, suffers a similar lack of systemic scrutiny and address. That is, if no one raises a particular issue, then that issue or concern may evade attention and address. Additionally, both ELSI and RRI are frequently the domain of ethicists and philosophers rather than public health and health policy scholars, which affects perspective as well as objectives and grounding. As such, pathways to health inequities may receive insufficient and inconsistent attention and address not commensurate with the public health crisis that it has become.

Two axis of technological innovation are useful in terms of exploring technological pathways to health inequities – 1) technological innovation pipeline or the phases of translation and 2) regulatory modalities particularly as applied to technology. The phases of technology development are fairly standard in concept although they may vary in label and parameters. They can be generally described as 1) idea or discovery, 2) proof of concept, 3) trials (clinical or other), 4) policy translation (including FDA or EMA approval), 5) uptake and integration, and 6) post-market monitoring (see e.g. Makni et al. 2017).

These stages can be viewed as potential points of intervention and, indeed, often operate in this way, although not necessarily in a systematic, coherent, or purposeful manner. Instead, clinical trials are evaluated, including for appropriateness of study sample, while other potential pathways and foundational research, and end-stage translation are subject to their own systems of scrutiny (see Char et al. 2020), which may or may not prioritize scrutiny for potential pathways to health inequities. Translation is, of course, a very broad concept, sometimes used to refer to the entire process, as in "from bench to bedside". Here, for the sake of convenience and brevity, I refer separately to these elements of translation. A great deal can be said about how each of these stages might present opportunities to intervene, however here only a brief reference will be made simply to illustrate possible ways of viewing the technological development trajectory with regard to intervention in pathways to health inequities.

6.1 Idea

At the formulation of an idea for tech innovation in health care, it may be difficult to foresee how the idea may unfold to privilege some and disadvantage others. Nevertheless, factors such as cost and potential applications (including potential misuse, e.g. research operating as scientific racism) may suggest an inequitable downstream impact even at this early stage. Among the more problematic aspects of dealing with technological pathways at the idea stage are: 1) a culture supporting scientific pursuit and 2) an inconsistent need-based orientation toward technological development. This second aspect is particularly concerning for health equity in that avoidable differences in health outcomes that consistently indicate a need for redress are ignored in favor of other, perhaps more profitable technological development actually serves to perpetuate disparities simply by failing to prioritize the ways that they might be addressed. For example, if diabetes and cardiovascular disease show a high and increasing incidence among the general population, yet ideas gravitate around technological innovation for diseases of lower prevalence diseases or only affordable to "elite" segments of the population, it becomes clear that even the idea stage hosts pathways to health inequities. A particularly compelling example is found in technological innovation in the treatment of Alzheimer's disease. Sadly, despite decades of scientific advances in the field, there is still no effective cure. Yet, much of the research effort focuses on hi-tech early detection technologies and prohibitively expensive treatments, apparently to the neglect of inexpensive interventions, with the predictable downstream effects (Pierce 2019). Various aspects of technological development and innovation are susceptible to serving as pathways to health inequities.

A highly concerning practice includes the marginalization of low cost interventions. For example, the incidence of Alzheimer's disease is increasing steadily, and has shown a pronounced increase in low- and middle- income countries. Yet, the primary avenues of research are awash in high technology elements (Pierce 2018). Although Alzheimer's research has increasingly devoted resources to investigations pertaining to the disproportionate incidence in African American communities, technological pathways in other endeavors should not be overlooked.

6.2 Proof of concept/funding

The selection of research for innovation has been recognized as a potential pathway to health inequities. Although proof of concept is, rightly, primarily a showing that the innovative mechanism can work, it is worth asking whether there is sufficient scrutiny of potential pathways to inequity at this stage from funders (particularly government). An issue that is increasingly raised in this regard is that of funding priorities, noting that research on health inequities has received relatively less funding (Carnethon et al. 2020).

6.3 (Clinical) Trials

As noted earlier, the limitations on generalizability and the problem of knowledge and understanding "not" gained as a result of lack of inclusion of persons other than white males in clinical research are well-documented (see e. g. Epstein 2008). Non-diverse study samples are not the only potential pathways to health inequity in research involving humans. Other factors such as stigma, study design and methods (e. g. frequency or duration of participation demands), exclusion and inclusion criteria, and site of participation may create avenues through which health inequities may unfold in the course of clinical research. All of these, in principle, may present as opportunities to interrupt pathways to inequities by giving deliberate attention to the ways that these features may play out against a backdrop of structural racism and differential economic status.

6.4 Translation

In principle, the general idea of the broad category of translation is designed to craft and guide optimal integration of innovation into practice. This includes a wide range of considerations ranging from price-setting, setting criteria for access, identifying and circumscribing intended use, identifying addressing ethical and legal issues and making adaptations accordingly. Clearly, this stage demands consideration of possible pathways to inequity. One key consideration in the translational phase (not necessarily a part of the component stages mentioned above) is that of access, a relatively familiar pathway to health inequities that can sometimes be so convoluted that it is mistaken for a neutral or coincidental cause. For example, a study in Denmark showed that "non-whites are 4.3 percentage points less likely than whites to receive special dementia care" (Stevnsborg 2016, p. 508). In this case, the access pathway to health inequities followed a regional resource allocation route that by virtue of the location of nursing homes that hosted predominantly white versus predominantly non-white dwellers. The allocation of technological equipment needed to administer the specific type of care was based on region, which, against a backdrop of structural racism, resulted in health inequities (see Sengupta 2012).

Indeed, ELSI is often considered to be the vehicle that is used to address downstream impacts that may be problematic but, as noted, the robustness of these inquiries is dependent on ad hoc identification of potential issues, thereby lacking the systematic and dedicated attention that eradication of health inequities requires. This may be influenced by perspective as well as discipline, in which case, diverse perspectives would logically be more likely to draw out a broader spectrum of impacts (and pathways). For this reason, the translational phase would be well served by the introduction of a more systematic approach to identifying and interrupting pathways to health inequities. One promising approach is the use of a health equity impact assessment (see e.g. Buse et al. 2019). Like the concept precursor, environmental impact assessments, depending on how it is crafted, health equity impact assessments aim to provide a more systematic approach to identify pathways of inequity and, at the same time, offer a dedicated, deep dive into the possible impacts of structural racism with regard to a particular innovation.

6.5 Post-market monitoring/Loop back into research

The value of using "use data" to inform further development and refinement of innovation is generally not disputed. However, this phase also presents a further opportunity for scrutiny regarding pathways to inequity of recent innovations. A deliberate solicitation of data pertaining to the experience, use, access, and outcomes by racial and ethnic minorities, different SES groups, or other categories of marginalized people, would, in theory, provide a critical window on what may, over time, evolve into significant pathways to health inequity. As has been recently observed, the importance of collection of data by such categories is likely to prove essential to identifying not only impacts (Flinagan et al. 2021), but also possible avenues of address.

7 Regulating for equity?

Among the approaches that have been deployed in the effort to eradicate racism as a causative factor is education (White-Davis 2018). Medical education of health personnel in all roles has increasingly emphasized the role of physicians, nurses, counselors, and other health personnel in contributing to differential health outcomes for different groups of people. This is essential, especially in view of the prevalence of implicit bias, resulting in unintended problematic behavior. However, education as an intervention strategy is based on a sort of "deficit model" of "undesirable" behavior. The deficit model of changing behavior generally refers to "educational" approaches that stem from a belief that if individuals had more or the right information, they would change their behavior (Bak 2001). As has been repeatedly shown across domains (e.g. genetics education regarding the merit of a link between genetics and race (Donovan 2020) or smoking (Baldwin et al. 2006), this is not necessarily the case. Other factors can override knowledge. In fact, some scholars in race and psychology summarize the phenomenon by the statement that "culture trumps everything" (Menachem 2018). Thus, the presence of cultural racism is constantly at work regardless of targeted efforts to change specific behaviors. With a view that a deficit model approach may address some issues, it may be insufficient to address such a complex and pervasive problem as structural and cultural racism, and could be well served by complementary and parallel approaches that operate in complementary ways. Just as employment law and housing laws have been enacted to counter, hinder, and reduce the effects of undesirable behavior (rather than the attitudes that are at the root of problematic behavior), health care delivery and health policy also have a range of options in the crafting of interventions. Perhaps one area that may be underutilized in strategies to eradicate health inequities is the multiple dimensions of regulation.

Regulation theory emphasizes that "regulation" is not only about the crafting of laws and penalties, but rather regulation refers to "any sustained attempt to influence behavior" (see Black 2002). Legal scholar Lawrence Lessig has identified four modalities of regulation – 1) laws, 2) market (economic incentives and disincentives), 3) social norms (social conventions that one feels compelled to follow), and 4) code (architecture; technological design to facilitate or inhibit certain behavior) (Lessig 1999). The field of public health has utilized social norms, for example, to powerful effect, for example, in the campaign against smoking (Karasek 2012). The point is to reinforce and complement ongoing strategies with systematic attention to interrupting technological pathways to health inequities, and regarding regulatory modalities deployed to regulate technology generally, but also to apply it specifically in developing strategies to eliminate health disparities. Regulation theory, in the broad sense of "sustained attempts to influence behavior" arguably offers a complementary way of approaching the work of eradicating health inequities. In the following section, I briefly outline how regulatory modalities may provide a useful platform for crafting or refining effective means of interrupting pathways to health inequities.

7.1 Regulatory modalities

Regulatory modalities of law, the market, social norms, and code are essentially strategies to influence or change behavior in a sustained way. When policy-makers perceive that a particular behavior needs to change, e.g. for the benefit of public health, there are many options available to achieve that goal. For example, as obesity rates became increasingly alarming, policy-makers drew multiple strategies from the regulatory toolbox and enacted such interventions as taxes on sugary products (market modality), banning of super-size drinks (law), encouraged or required food/caloric labelling and re-ordered shelves of healthy and non-healthy foods (nudge or code), and education (building toward social norms). These approaches could be used strategically in the effort to hinder and interrupt technological pathways to health inequities.

7.2 Law and regulation

Law and regulation are the most familiar regulatory tools, but these often carry disadvantages. In the context of health care delivery, there are laws prohibiting discrimination arising out of multiple regulatory schemes, e.g. health law, employment law, insurance law, and specific prohibitions such as the Genetic Information Non-Discrimination Act (GINA). However, the challenges of enforcement, intervening in a timely manner, and the post hoc nature of redress hinder broad based sustained reduction of the problem within relatively near-term horizons. Thus, while law serves many important functions, it is not the hammer for all nails. The market offers possibilities for changing behavior in the appropriate and strategic creation of economic incentives and disincentives that may interrupt technological pathways. Thus, a system of subsidies, taxes, or other economic advantages or disadvantages could be used to incentivize (and disincentivize) equity-promoting behavior in the use of technology in healthcare. Social norms, which some regard as perhaps the strongest of regulatory modalities in that it relies of informal social censure as a way of regulating behavior and operates pervasively in one's environment could be a powerful tool. Monitoring and enforcement is built into the relevant community and the "penalty" for non-compliance with social norms can be quite rapid and robust in the social exclusion or other forms of informal community condemnation that can ensue. Smoking and littering are two well-known examples of two successful campaigns to shift social norms (Sunstein 2014). Of course, social norms carry important disadvantages, such as legitimacy, that should inform how such campaigns are crafted, by whom, with which social penalties, and to achieve what kinds of ends. Interesting examples of the use of a form of social norms to influence behavior in technological development can be found in the field of standardization. Many standards are not required by law but arise through a consensus process that agrees on objectives and what may be required to achieve them. The use of social norms to influence behavior in efforts to eradicate the causes of health inequities, particularly racism, has been a long-standing approach, but generally is without a coherent overarching strategy and is largely dependent on the robustness of the commitment of the relevant community. Education efforts in medical education take an important step in this direction and could be supplemented by other mechanisms that "strengthen" the impact of education. Strengthening social norms that condemn biased behavior in healthcare delivery that result in social consequences (in the relevant community, e. g. hospital, professional association) may be worthy of consideration.

7.3 Code

Lastly, code is perhaps the most familiar regulatory tool in the field of technological development. AI-driven technologies, for example, can lend themselves to subject to design requirements aimed at ensuring equity-promoting use without constant human oversight.

Code is a common approach to ensuring that technology does not operate in ways contrary to certain values is to build the values into the design. Nudges, a form of choice architecture that steers but does not compel behavior, can also fall into the category of code, and often operates as a powerful incentive to perform or refrain from certain behaviors (Sunstein 2014). The relatively recent flood of critiques about AI perpetuating inequities in healthcare delivery and elsewhere has pointed to design as a possible way of curtailing this pathway to inequities, requiring certain types of training data, excluding the use of certain characteristics in labelling of inputs to calculate outputs that are intended to inform decision-making. Other strategies such as requiring a degree of transparency that allows for necessary scrutiny of possible bias in the generation of decision-informing outputs. The excellent book "Automating Inequality" by Virginia Eubanks provides a relentless and compelling view of how AI technology serves as a profoundly powerful conduit to inequities, health and otherwise (Eubanks 2018). Drawing attention to the urgent need to develop ways to ensure that the use of AI technologies in healthcare does not perpetuate or increase health inequities (see e. g. Vyas 2020), Eubanks has been joined by a growing call for intensified scrutiny of AI in healthcare with regard to bias and inequitable impacts on health outcomes.

8 Looking forward

Technological innovation in healthcare requires scrutiny, not just for safety, but also for its deleterious social effects, and ultimately disparate impacts on peoples' lives and well-being in the form of health inequities. While this chapter has highlighted digital technologies because of its pervasiveness and increasing role in clinical care and public health, technological innovation in health in genetics, cardiology, nephrology, neurology, pain management, to name only a few, merit robust and systematic scrutiny for the creation and perpetuation of pathways to health inequities. While the critically important work of documenting health inequities continues, the ongoing scale and nature of health inequities along racial, ethnic, and SES is inexcusable and, indeed, increasingly acknowledged to be "a public health crisis". Worse, it is an avoidable one. Medical anthropologist Arthur Kleinman has declared that the background realities of the lived experience must be brought into view and that "we cannot *not* scrutinize implications for care, policy and research in medical education, medical practice, and medical research" (Kleinman 2012, p. 1551; see also Katz 2020). As we "scale up" technologically and become increasing reliant on technology in the delivery of healthcare, it is imperative to ask whether the increasing technological sophistication and digitization is serving everyone equally or is technology-facilitated healthcare ultimately serving to exacerbate, perpetuate, and even create health inequities? This chapter insists that we also scale up deliberate regard for interrupting technological pathways to health inequities through systematic and timely intervention, making use not only of the technology development trajectory, but also the regulatory toolbox that has as its goal to engage in sustained attempts to influence behavior.

References

Anderson, Karen O,; Green, Carmen R.; Payne, Richard. (2009): "Racial and ethnic disparities in pain: causes and consequences of unequal care". In: *The Journal of Pain* 10. N. 12, pp. 1187–1204.

Bak, Hee-Je. (2001): "Education and public attitudes toward science: Implications for the "deficit model" of education and support for science and technology". In: *Social Science Quarterly* 82. N. 4, pp. 779–795.

Baldwin, Austin S.; Rothman, Alexander J.; Hertel, Andrew W.; Linde, Jennifer A.; Jeffery, Robert W.; Finch, Emily A.; Lando, Harry A. (2006): "Specifying the determinants of the initiation and maintenance of behavior change: an examination of self-efficacy, satisfaction, and smoking cessation". In: *Health Psychology* 25. N. 5, 626.

Beckwith, Jonhatan; Pierce, Robin. (2018): "Genes and Human Behavior". In Gerlai, Robert T. (Ed), *Molecular-Genetic and Statistical Techniques for Behavioral and Neural Research*, pp. 599–622. London, Boston, Oxford, New York, Sand Diego: Academic Press.

Beckwith, Jonhatan; Bergman, Kostia; Carson, Michael; et al. (2017): "Using dialogues to explore genetics, ancestry, and race". In: *The American Biology Teacher* 79. N. 7, pp. 525–537.

Benjamin, Ruha. (2019): *Race After Technology*. London: Polity Press.

Black, Julia. (2002): "Critical Reflections on Regulation". In: *Australian Journal of Legal Philosophy* 27. Pp. 1–36.

Braveman, Paula. (2014): "What are health disparities and health equity? We need to be clear". In: *Public health reports* 129. Suppl. 1, pp. 5–8.

Braun, Lundy. (2014): *Breathing race into the machine: The surprising career of the spirometer from plantation to genetics*. Minneapolis: University of Minnesota Press.

Buse, Chris; Lai, Valierie; Cornish, Katie; Parkes, Margot W. (2019): "Towards environmental health equity in health impact assessment: innovations and opportunities". In: *International journal of public health* 64. N. 1, pp. 15–26.

Carnethon, Mercedes R.; Kershaw, Kiari; Kandula, Narmatha R. (2020): "Disparities research, disparities researchers, and health equity". In: *Jama* 323. N. 3, pp. 211–212.

Chang, Virginia W.; Lauderdale Diane S. (2009): "Fundamental Cause Theory, Technological Innovation, and Health Disparities: The Case of Cholesterol in the Era of Statins". In: *J Health Soc Behav* 50. N. 3, pp. 245–260.

Char, Danton. S.; Abràmoff, Michael D.; Feudtner, Chris. (2020): "Identifying ethical considerations for machine learning healthcare applications". In *AJOB* 20. N. 11, pp. 7–17.

Collins, Francis. S.; Patrinos, Ari; Jordan, Elke; et al. (1998): "New goals for the US human genome project: 1998–2003". In: *Science* 282. N. 5389, pp. 682–689.

Cuevas, Adolfo G.; Chen, Ruijja; Slopen, Natalie; et al. (2020): "Assessing the role of health behaviors, socioeconomic status, and cumulative stress for racial/ethnic disparities in obesity". In: *Obesity* 28. N. 1, pp. 161–170.

Donovan, Brian M.; Weindling, Monic; Salazar, Brae, et al. (2021): "Genomics literacy matters: Supporting the development of genomics literacy through genetics education could reduce the prevalence of genetic essentialism". In: *Journal of Research in Science Teaching* 58. N. 4, pp. 520–550.

Duster, Troy. (2004). *Backdoor to eugenics*. London: Routledge.

Elster, Jon. (1989): "Social norms and economic theory". In: *Journal of economic perspectives* 3. N. 4, pp. 99–117.

Epstein, Steven. (2008): *Inclusion: The politics of difference in medical research*. Chicago: University of Chicago Press.

Eubanks, Virginia. (2018): *Automating inequality: How high-tech tools profile, police, and punish the poor*. New York: St. Martin's Press.

Flanagin, Annette; Frey, Tracy; Christiansen, Stacy L.; et al. (2021): "AMA Manual of Style Committee. Updated Guidance on the Reporting of Race and Ethnicity in Medical and Science Journals". In: *JAMA* 326. N. 7, pp. 621–627. doi:10.1001/jama.2021.13304.

Gates, Henry Louis. (2020): *Stony the Road: Reconstruction, White Supremacy, and the Rise of Jim Crow*. London: Penguin Books.

Geiger, Jack. (2003): "Racial and ethnic disparities in diagnosis and treatment: a review of the evidence and a consideration of causes". In: Smedley, Brian D.; Stith, Adrienne Y.; Nelson, Alan R.; et al.; *Unequal treatment: Confronting racial and ethnic disparities in health care*. https://www.ncbi.nlm.nih.gov/books/NBK220337/, visited on 22 July 2022.

Glied, Sherry; Lleras-Muney, Adriana. (2008): "Technological innovation and inequality in health". In: *Demography* 45. N. 741, pp.741–761.

Goldman, Dana P.; Lakdawalla, Darius N. (2005): "A theory of health disparities and medical technology". In: *Contributions in Economic Analysis & Policy* 4. N. 1, pp. 1–30.

Katz, Arlene M. (2020): "Social poetics as processual engagement: Making visible what matters in social suffering". In: *Transcultural Psychiatry* 20. doi: 10.1177/1363461520962614.

Kleinman, Arthur. (2012): "Caregiving as moral experience". In: *The Lancet* 380. N. 9853, pp. 1550–1551.

Kumanyika, Shiriki. (2012): "Health disparities research in global perspective: new insights and new directions". In: *Annual review of public health* 33. Pp. 1–5.

Karasek, Deborah; Ahern, Jennifer; Galea, Sandro. (2012): "Social norms, collective efficacy, and smoking cessation in urban neighbourhoods". In: *American journal of public health* 102. N. 2, pp. 343–351.

Katell, Michael; Young, Meg; Dailey, Dharma; et al. (2020): "Toward situated interventions for algorithmic equity: lessons from the field". In: *Proceedings of the 2020 conference on fairness, accountability, and transparency*. Pp. 45–55. https://dl.acm.org/doi/abs/10.1145/3351095.3372874, visited on 22 July 2022.

Kennel, Jamie; Withers, Elisabeth; Parsons, Nate; Woo, Hyeyoung. (2019): "Racial/ethnic disparities in pain treatment: evidence from Oregon emergency medical services agencies". In: *Medical care* 57. N. 12, pp. 924–929.

King Jr, Talmadge E. (2002): "Racial disparities in clinical trials". In: *N Engl J Med* 346. Pp. 1400–1402.

Koenecke, Allison; Nam, Andrew; Lake, Emily; et al. (2020): "Racial disparities in automated speech recognition". In: *Proceedings of the National Academy of Sciences* 117. N. 14, pp. 7684–7689.

Korda, Rosemary J.; Clements Mark S.; Dixon, Jane. (2011): "Socioeconomic inequalities in the diffusion of health technology: Uptake of coronary procedures as an example". In. *Soc Sci Med* 72. N. 2, pp. 224–229.

Krieger, Nancy. (2021): "Structural racism, health inequities, and the two-edged sword of data: structural problems require structural solutions". In: *Frontiers in public health* 9. N. 655447. doi: 10.3389/fpubh.2021.655447.

Lafontaine, Alika. (2018): "Indigenous health disparities: A challenge and an opportunity". In: *Canadian Journal of Surgery* 61. N. 5, 300–301.

Lessig, Lawrence. (1999): *Code and Other Laws of Cyberspace.* New York: Basic Books.

Lutfey, Karen; Freese, James. (2005): "Toward some fundamentals of fundamental causality: socioeconomic status and health in the routine clinic visit for diabetes1". In: *American Journal of Sociology* 110. N. 5, pp. 1326–1372.

Mackenbach Johan P. (2012): "The persistence of health inequalities in modern welfare states: the explanation of a paradox". In: *Soc Sci Med* 75. Pp. 761–769.

Makhni, Sonia; Atreja, Ashish; Sheon, Amy; et al. (2017): "The broken health information technology innovation pipeline: a perspective from the NODE health consortium". In. *Digital biomarkers* 1. N. 1, pp. 64–72.

Malmusi, Davide. (2015): "Immigrants' health and health inequality by type of integration policies in European countries". In: *The European Journal of Public Health* 25. N. 2, pp. 293–299.

Menachem, Resmaa. (2021): *My Grandmother's Hands: Racialized Trauma and the Pathway to Mending Our Hearts and Bodies.* London: Penguin Books.

McGowan, Angela K.; Lee, Mary M., Meneses, Cristina; et al. (2016): "Civil rights laws as tools to advance health in the twenty-first century". In: *Annual Review of Public Health* 37. Pp. 185–204.

Moran, Rachel F. (2020): "Persistent Inequalities, the Pandemic, and the Opportunity to Compete". In: *Wash & Lee J Civ Rts & Soc Just* 27. N. 2, pp. 589–644.

Morris, Corey; Shen, Aimee; Pierce, Robin; Beckwith, Jonathan. (2013): "Deconstructing violence". In: Krimsky, Sheldon; Gruber, Jonathan (eds), *Biotechnology in Our Lives,* pp. 233–244. New York: Skyhorse Publishing.

Obermeyer, Ziad; Powers, Brian; Vogeli, Christine; & Mullainathan, Sendhil. (2019): "Dissecting racial bias in an algorithm used to manage the health of populations". In: *Science* 366. N. 6464, pp. 447–453.

Owen, Richard; Macnaghten, Phil; Stilgoe, Jack. (2012): "Responsible research and innovation: From science in society to science for society, with society". In: *Science and public policy* 39. N. 6, pp. 751–760.

Parikh, Ravi B.; Obermeyer, Ziad; Navathe, Amol S. (2019): "Regulation of predictive analytics in medicine". In: *Science* 363. N. 6429, pp. 810–812.

Pierce, Robin (2018); "Challenges for Collective Approaches to Alzheimer's Disease Research". In: Beers, Britta; Sterckx, Sigrid; Dickenson, Donna; (Eds.), *Personalised Medicine, Individual Choice, and the Common Good.* Cambridge, Massachusetts: Cambridge University Press.

Pierce, Robin. (2018): "Machine learning for diagnosis and treatment: Gymnastics for the GDPR". In: *European Data Protection Law Review* 4. N. 3, pp. 333–343.

Pierce, Robin; Sterckx, Sigrid; Van Biesen, Wim. (2021): "A riddle, wrapped in a mystery, inside an enigma: How semantic black boxes and opaque artificial intelligence confuse medical decision-making". In: *Bioethics* 26. N. 2, pp. 113–120.

Piot, Peter. (2012): "Innovation and technology for global public health". In: *Global public health* 7. Suppl. 1, pp. S46-S53. https://doi.org/10.1080/17441692.2012.698294.

Pirisi, Angela. (2015): "Country in Focus: health disparities in Indigenous Canadians". In: *The Lancet Diabetes & Endocrinology* 3. N. 5, pp. 319.

Schmidt, Harald; Gostin, Lawrence O.; Williams, Michelle A. (2020): "Is it lawful and ethical to prioritize racial minorities for COVID-19 vaccines?". In: *Jama* 324. N. 20, pp. 2023–2024.

Shonkoff, Jack P.; Slopen, Natalie; Williams, David R. (2021): "Early childhood adversity, toxic stress, and the impacts of racism on the foundations of health". In: *Annual Review of Public Health* 1. N. 42, pp. 115–134.

Smith, Caren E.; Fullerton, Stephanie M.; Dookeran, Keith A.; et al. (2016): "Using genetic technologies to reduce, rather than widen, health disparities". In: *Health Affairs* 35. N. 8, pp. 1367–1373.

Sturgis, Patrick; Allum, Nick. (2004): "Science in society: re-evaluating the deficit model of public attitudes". In: *Public understanding of science* 13. N. 1, pp. 55–74.

Sunstein, Cass R. (2014): *Why nudge?*. Boston, Massachusetts: Yale University Press.

Sunstein, Cass R. (1996): *Social norms and social roles*. In: *Chicago Law & Economics Working Paper* 36. https://ssrn.com/abstract=10001, visited on 12 July 2022.

Tehranifar, Parisa; Neugut, Alfred I.; Phelan, Jo. C.; et al. (2009): "Medical advances and racial/ethnic disparities in cancer survival". In: *Cancer Epidemiology and Prevention Biomarkers* 18. N. 10, pp. 2701–2708.

Vo, Jacqueline B.; Gillman, Arielle; Mitchell, Kelsey; Timiya, S. Nolan. (2021): "Health Disparities: Impact of Health Disparities and Treatment Decision-Making Biases on Cancer Adverse Effects Among Black Cancer Survivors". In: *Clinical Journal of Oncology Nursing*, 25. N. 5, pp. 17–24.

Vyas, Darshali A.; Eisenstein, Leo; Jones, David S. (2020): "Hidden in plain sight—reconsidering the use of race correction in clinical algorithms". In: *New England Journal of Medicine* 383. N. 9, pp. 874–882.

Wachter, S. (2020): "Affinity Profiling and Discrimination by Association in Online Behavioral Advertising". In: *Berkeley Tech* 35. N. 2, 367. https://papers.ssrn.com/sol3/papers.cfm?abstract_id=3388639, visited on 22 July 2022.

Weiss, Daniel; Rydland, Håvard T.; Øversveen, Emil; et al. (2018): "Innovative technologies and social inequalities in health: a scoping review of the literature". In: *Plos One* 13. N. 4, e0195447.

White, Douglas B.; Lo, Bernard. (2021): "Mitigating inequities and saving lives with ICU triage during the COVID-19 pandemic". In: *American Journal of Respiratory and Critical Care Medicine* 203. N. 3, pp. 287–295.

White-Davis, Tanya; Edgoose, Jennifer; Speights, Joedrecka B.; et al. (2018): "Addressing racism in medical education an interactive training module". In: *Family medicine* 50. N. 5, pp. 364–368.

Wilkerson, Isabel. (2020) *Caste*. London: Penguin Books.

Williams, David R.; Collins, Chiquita. (2001): "Racial residential segregation: a fundamental cause of racial disparities in health". In: *Public Health Reports* 116. N. 5, pp. 404–416.

Williams, David R.; & Wyatt, Ronald. (2015): "Racial bias in health care and health: challenges and opportunities". In. *JAMA* 314. N. 6, pp. 555–556.

Williams, David R.; Lawrence, Jourdyn A.; Davis, Brigette A. (2019): "Racism and health: evidence and needed research". In: *Annual Review of Public Health* 40. Pp. 105–125.

Williams, David R. (2012): "Miles to go before we sleep: Racial inequities in health". In: *Journal of Health and Social Behavior* 53. N. 3, pp. 279–295.

Williams, David R.; Mohammed, Selina A. (2009): "Discrimination and racial disparities in health: evidence and needed research". In. *Journal of Behavioral Medicine* 32. N. 1, pp. 20–47.

Woolf, Steven; Johnson, Robert; Phillips, Robert; et al. (2007): "Giving everyone the health of the educated: An examination of whether social change would save more lives than medical advances". In: *Am J Public Health* 97. N. 4, pp. 679–683.

Yancy, Clyde W. (2020): "*COVID-19 and African Americans*". In: *JAMA* 323. N. 19, pp. 1891–189

Célia Mariana Barbosa de Souza, Fernanda Sales Luiz Vianna,
Francisco Veríssimo Veronese

Investigating the Relationship between Chronic Kidney Disease and the Black Population

Abstract: Brazil was the largest slave-keeping territory in the Western Hemisphere, and the lack of governmental policies benefitting African-descendants as well as fantasies regarding their inferiority contributed to the continuity of racism in this country. Even after the abolition of slavery, the Brazilian government perpetuated racist behaviours by attracting European immigrants with the goal of "whitening" the population. These historical events not only have socioeconomic and cultural impact but also influence the health of the black population. Chronic kidney disease (CKD) is a public health problem that can lead to a decrease in kidney function and necessitate dialysis treatment in all populations. However, this problem is more concerning in people of African heritage since it is more prevalent in black populations. Recent studies have associated CKD with the genetic background of African populations by linking it to genetic variants related to protection against sleeping sickness, which is endemic in Africa. Thus, it is fundamental to emphasize the importance of studies investigating populations of African descent or featuring African admixture because they can improve strategies for CDK prevention and public health policies aimed at the black population. These studies have illustrated the need for specific public actions regarding particular populations, especially those that have been marginalized by secular processes of exclusion and subjugation.

1 Introduction

Brazil is the largest country in Latin America, with an area of 8,510,345,538 km² and a population projection of 214,266,424 inhabitants (IBGE 2022). The arrival of African captives in Brazil began in the 15th century, in approximately 1553, based on the transatlantic traffic of black men and women who were enslaved to sustain colonial expansion. These individuals were forced to work in sugar mills and

Célia Mariana Barbosa de Souza, Federal University of Rio Grande do Sul
Fernanda Sales Luiz Vianna, Federal University of Rio Grande do Sul
Francisco Veríssimo Veronese, Federal University of Rio Grande do Sul

https://doi.org/10.1515/9783110765120-006

on sugarcane plantations. The trafficking of black men and women was facilitated by slave ships. Many Africans died on the way due to poor facilities, food deprivation and violence suffered during the journey (Munanga 2009).

Brazil was the largest slave-keeping territory in the Western Hemisphere for nearly three and a half centuries. It received approximately 5 million African captives, 40% of the total of 12.5 million such captives shipped to America (Gomes 2019). The slaves trafficked to Brazil came from Mozambique, Angola, Congo and the coast of Central Africa to Bahia; from the Bay of Benin, the coast of Namibia and Senegambia to Pernambuco and Maranhão; and from Mozambique to the southeast of Brazil (Rio de Janeiro and São Paulo) (Gomes 2019).

Slavery was an enormous humanitarian tragedy in Brazil. Torn away from the continent and culture in which they were born, Africans and their descendants built Brazil with their hard work. They were humiliated, exploited, discriminated against and subjected to violence. This experience was the most defining experience in Brazilian history, with a profound impact on the culture and political system that gave birth to the country following its independence in 1822 (Gomes 2019).

The official abolition of slavery took place in 1888 during the Brazilian Empire (1822–1889), after the black population had experienced slavery for more than three centuries. Despite abolition and the subsequent acknowledgement of the serious violations inherent in slavery, the black population remains largely socially excluded from access to many goods and the enjoyment of fundamental rights (Brazil 2016).

The lack of political policies in the post-abolition period that included this population in productive and social processes as well as fantasies regarding their inferiority contributed to the continuation of their marginalization in Brazilian society (Brazil 2016).

At present, racial discrimination continues to function as a barrier to an egalitarian society, creating competition for social opportunities between blacks and whites while guaranteeing a competitive advantage to the white population, which is considered to be "superior" (Carneiro 2011).

2 Understanding Racism

The Swedish naturalist Carl Linnaeus (1707–1778), in the 1767 edition of his *Systema Naturae*, presented a taxonomic division of the human species, distinguishing four main races and qualifying them according to what he believed were their main characteristics: *Homo sapiens Europaeus*, which would be white, serious, strong; *Homo sapiens Asiaticus*, which would be yellow, melancholic, avaricious; *Homo sapiens Africanus*, which would be black, impassive, lazy; *Homo sapiens Americanus*, which would be red, ill-tempered, violent (Pena 2008).

The French philosopher Voltaire (François-Marie Arouet 1694–1778), a contemporary of Linnaeus, stated in his *Philosophical Letters*, which were published in 1733: "The negro race is a species of men as different from ours as the breed of spaniels is from that of greyhounds [...] The black wool on their heads and other parts has no resemblance to our hair; and it may be said that if their understanding is not of a different nature from ours, it is at least greatly inferior". Considering both lines of thought, scientific racism was therefore reinforced and legitimated (Pena 2008).

Racism is a phenomenon whose dynamics are rooted in the structures of society. The meanings it reproduces affect the treatment given to various populations, influencing access to health, education, housing and opportunities. Therefore, racism creates and/or strengthens vulnerability by imposing barriers and neglecting needs (Brazil 2016).

Racism is a social determinant of health, since health and illness are related to various socioeconomic and cultural factors that affect the physical and the psychological as well as both individual and collective integrity. The historical conditions of social insertion, alongside housing, income, health, geographical location and positive or negative self-concept, are also elements that determine access to health goods and services (Brazil 2016).

The centuries of enslavement experienced by the black population influenced negatively the inclusion of this population in Brazilian society, contributing to an unequal and unfavorable access to rights and opportunities, including those related to health. These characteristics are reflected in the epidemiologic profile of this population, which exhibits inequities and vulnerabilities with respect to health-promoting conditions (Brazil 2016). This process did not occur without resistance: the *quilombos*[1] are one example of a strategy of organization and opposition to the slave system in Brazil (Munanga 2004).

The official source for data concerning the racial composition of the population of Brazil is the Brazilian Institute of Geography and Statistics (IBGE), which is responsible for the demographic census and constitutes the main reference for understanding the living conditions of the population in all municipalities of the country. The parameter that IBGE uses to classifying the population (which is done on a decennial basis) is the "color/ethnicity question". The ques-

---

[1] *Quilombo* was an organization of black people who fled from slavery to a place that was difficult to access by the plantation masters and implemented a socio-political structure (see Thornton 2010).

tion posed to citizens during the census allows each individual to answer according to five categories: white, black, brown (better known as *pardo*[2]), yellow and indigenous (IBGE 2022).

Brazil has 26 national states in addition to the Federal District, where Brasília, the capital of the country, is located. Since 1970, Brazil has been divided into the following regions[3]: North, Northeast, Southeast, South and Central-West (IBGE 2022).

The results of the National Survey by Continuous Household Sample (PNAD) 2012–2019, which was conducted by IBGE, indicated the total percentage of Brazil's white and black[4] populations (46.6% white and 56.2% black), as well as the following percentages for each region in 2019: North Region: 19.1% white and 79.5% black; Northeast Region: 24.7% white and 74.4% black; Southeast Region: 50.5% white and 48.9% black; South Region: 73,2% white and 25.9% black; and Central-West Region: 36.2% white and 62.6% black (IBGE 2022).

The decreased presence of the black population in the states of Rio Grande do Sul, Santa Catarina and Paraná – compared to the other states in Brazil –, is due to the migration of Africans towards other regions of Brazil. After the abolition of slavery, the Brazilian government implemented racist actions with the objective of "whitening" the Brazilian population, which resulted in a law encouraging European immigrants to settle in the country. In 1884, Law No. 28, which was approved by the São Paulo State legislature, guaranteed resources to allow the state government to finance immigration and noted that the beneficiaries of these resources were to be European workers and their families:

> Authorizes the government to support immigrants from Europe and Azores and Canary Islands, to settle in the province, with the following financial support: 70,000$ for those over 12, 35,000$ for those from 7 to 12, and 17,500$ for those from 3 to 7 years of age" (Law No. 28, 03/29/1884).

The black population has historically been located on the bottom of the social pyramid, as indicated by the fact that out of 13,5 million Brazilians living in extreme poverty, 10,1 million declared themselves to be black or *pardo* (IBGE 2019).

---

2 *Pardo* refers to a person with an ancestry based on a mixture of skin colors, including whites, blacks, and indigenous people (Munanga 2004).
3 The regional division of Brazil consists of the grouping of states and municipalities into regions, which included the North Region, Northeast Region, Southeast Region, South Region, and Center-West Region; this division remains in effect at this time (IBGE 2022).
4 The black population in Brazil is defined in terms of the sum of the black and *Pardo* populations.

It is important to highlight the fact that, proportionally, there were fewer records of health assistance and hospitalization among blacks than among whites, indicating that the black population relies mostly on services offered by the *Sistema Único de Saúde* (SUS)[5], the Brazilian public health system (IBGE 2019).

The resistance of the black population against all forms of racism is currently organized in the Unified Black Movement (MNU). The MNU has pressured the Brazilian government to implement public policies that respect the rights of the black population. Throughout its activities, many seminars and campaigns were used to protest institutional racism, which led to the two important events:

1) The creation of the *Secretaria Nacional de Políticas de Promoção da Igualdade Racial* (National Secretariat for Policies Promoting Racial Equality) by Law No. 10.678/2003 (Brazil 2003).

2) The institution of the *Política Nacional de Saúde Integral da População Negra* (National Policy for the Integral Health of the Black Population, hereafter referred to as PNSIP) by the publication of Ordinance No. 992/2009 (Brazil 2009).

The PNSIP expresses a commitment on the part of the Ministry of Health to fight inequalities in the public health system and to promote the health of black population in an "integral way", i.e. by taking into account the fact that inequalities in health and healthcare are the result of unfair socioeconomic processes and cultural factors – including racism – that contribute to rates of morbidity and mortality among the Brazilian black population (Brazil 2009).

The implementation of this policy encourages managers, social movements, counselors and SUS professionals to work towards the improvement of health conditions among the black population. This includes developing an improved understanding of their vulnerabilities and striving to acknowledge the weight of racism as a social determinant of health (Brazil 2009).

The PNSIP includes more detailed objectives, such as guaranteeing and expanding access to healthcare for black people living both in urban and rural areas, particularly those living in peripheral regions and forests, including the *quilombola* settlements[6]; it also focuses on monitoring the indicators and targets established to promote the health of the black population, with the ultimately purpose of reducing regional and local inequalities (Brazil 2009). However,

---

5 The Brazilian SUS is one of the largest and most complex public health systems in the world and is composed of the Ministry of Health, states, and municipalities, as determined by the Federal Constitution of Brazil (Brazil 2021).

6 These settlements represented an important form of resistance to slavery when that institution was legal.

the policy has not achieved yet its objectives, mostly because its implementation depends on a difficult form of coordination among the three relevant administrative levels (federal, state and municipal).

3 Ancestry in Brazil

Due to its history as a colonized country, it is possible to argue that the current composition of the Brazilian population is the result of a series of complex instances of "miscegenation" or a process by which people from different groups become mixed (Naslavsky et al. 2017). Initially, the territory currently known as Brazil was populated by Native Americans, whose ancestors probably migrated from Asia via the Bering Strait (Gonçalves et al. 2013). In the 16th century, the European occupation of Brazil began with the arrival of the Portuguese and Spanish, who instituted the slave trade by bringing black Africans with them to Brazil (Naslavsky et al. 2017). This occupation, as a consequence, also led to a decline in the Native American population (Ruiz-Linares et al. 2014). Later, in the 19th century, new migration waves occurred, bringing individuals from other parts of Europe and Asia (Franco et al. 1999). Internal migration and the subjugation of Native American and black women by European men also contributed greatly to the genetic variety of the Brazilian population (Alves-Silva et al. 2000; Kehdy et al. 2015).

Brazil received approximately half of the total number of black slaves brought to America between the 16th and 19th century (Miller et al. 1997). It is estimated that more than 4 million Africans were brought to Brazil starting in approximately 1535 (Gomes 2019), and so Brazil was the largest recipient of slaves in all of the Americas (Ruiz-Linares et al. 2014). A meta-analysis of 25 studies examining 38 different populations in Brazil (De Moura et al. 2015) showed that the pooled contributions of European, African and Native American ancestry in Brazil were 0.62, 0.21 and 0.17, respectively. The South Region, which includes the states Rio Grande do Sul, Santa Catarina and Paraná, has the largest contingent of European descendants (0.77), while the Northeast Region has the largest proportion of African descendants (0.27) and the North Region has the largest percentage of Amerindians (0.32) (De Moura et al. 2015).

All these migration patterns make Brazil a unique country even compared to other Latin American countries (Ruiz-Linares et al. 2014), and this situation has resulted in a unique combination of rare and common genetic variants (Naslavsky et al. 2017).

Recent studies have shown a higher prevalence of genetic variants in the Apolipoprotein 1 (APOL1) gene in black people (Siemens et al. 2018; Riella et al. 2019), which is associated with a higher risk of developing CKD. These genetic variants associated with CKD are located in the gene called APOL1; this gene is responsible for producing the protein APOL1, which is involved in the transportation of lipids throughout the body. It is estimated that these genetic variants, called G1 and G2, emerged approximately 4000 years ago in Africa and subsequently increased rapidly in frequency (Friedman et al. 2016). Interestingly, this rapid increase in the frequency of such genetic variants has been associated with protection against trypanosomiasis (sleeping sickness), a very common condition in Africa (Friedman et al. 2011a; Freedman et al. 2021; Siemens et al. 2018). Studies have shown that approximately 46% of the black people in Nigeria and 36% of the black people in North America carry some of the risk variants of APOL1 associated with CKD (Friedman et al. 2011a; Freedman et al. 2021; Siemens et al. 2018). Despite providing protection against trypanosomiasis, the presence of the genetic variants entails a major risk of developing CKD.

Considering the fact that no studies have evaluated the occurrence of these genetic risk variants of APOL1 among the Brazilian population, the actions that promote and prevent CKD among black people are impaired, exposing these individuals to a faster and more prevalent progression of kidney disease to its final stages.

4 Chronic kidney disease in Brazil

CKD affects the health status of an individual on several levels and can lead both to a decrease in kidney function without clinical manifestations and to the development of end-stage renal disease (ESRD). The definition of CKD, according to Kidney Diseases Improving Global Outcomes (KDIGO) (Eknoyan et al. 2013), is the presence of kidney damage or decreased kidney function for more than three months alongside the related health implications; this process is classified into stages[7].

The Brazilian Society of Nephrology (SBN) plays a very important role in Brazil with respect to the dissemination of information, organization, and publication of data regarding CKD; its mission is to ensure universal access to kidney health care throughout society (SBN 2022).

---

7 The stages of CKD were defined in accordance with the level of the estimated glomerular filtration rate (eGFR) as follows: stage 1: >90 ml/min/1.73 m^2; stage 2: 60–89 ml/min/1.73 m^2; stage 3a: 45–59 ml/min/1.73 m^2; stage 3b: 30–44 ml/min/1.73 m^2; stage 4: 15–29 ml/min/1.73 m^2; stage 5: <15 ml/min/1.73 m^2.

In Brazil, the national dialysis registry published data concerning the period between 2000–2012, in which 40.5% of people in chronic dialysis program self-reported themselves to be black (i. e. black or *pardo*); nevertheless, 13.2% of dialysis patients provided no information regarding their color (Moura et al. 2014). The discrepancy between the frequency of CKD and treatment strategies for black people is the result of multiple factors, many of which are subject to modification. These factors include both modifiable factors (healthy diet, physical exercise, regular medical monitoring, frequent assessment of kidney function, and access to health care) and non-modifiable factors (genetics, age, color, and sex) (Moura et al. 2014).

In 2019, a study published in Brazil in conjunction with Harvard Medical School in Massachusetts, Boston (Riella et al. 2019) assessed the presence of the APOL1 risk genetic variants associated with kidney disease in Brazil in 106 individuals with African ancestry who were on hemodialysis; this group was compared to 106 healthy first-degree relatives (the controls) who lacked these variants. This study showed that the presence of the risk variants was 10 times more common in African descendants who were on hemodialysis than in the controls, and the carriers of both genetic variants started dialysis 12 years earlier than those who were not carriers of such genetic variants.

5 Chronic kidney disease in the United States of America

In 2021, the prevalence of End Stage Renal Disease (ESRD) in the United States of America (USA) was estimated to be significantly higher in African-American people than in white and Asian individuals, affecting 5000 African-American patients per million of the population in 2018 (Johansen et al. 2021).

In 2010, Genovese and colleagues (Genovese et al. 2010a) performed genetic mapping of African-Americans[8] in the US, identifying in 13% of individuals the genomic region of chromosome 22, which encompasses the APOL1 gene. APOL1 gene variants are common on the chromosomes of individuals of African descent but absent on the chromosomes of individuals of European descent (Friedman et al. 2011b).

---

8 African-American: a racial designation used by the U.S. Office of Management and Budget, the department that governs census reporting, which includes the following categories: white, American Indian or Alaska Native, Asian, and Black or African-American (Norris et al. 2017).

According to Saran (Saran et al. 2016), terminal CKD (ESRD) is one of the most vivid examples of health indicator disparities in the USA among the African-American population. In this context, it is noted that although diabetes mellitus and hypertension are the leading causes of ESRD, in most cases, the influence of socioeconomic status, lifestyle, and other clinical factors may contribute to the higher risk of ESRD among African-Americans.

There is a significant disparity in the frequency of segmental and focal glomerulosclerosis (GESF) between African-Americans and European descendants; the former have four times the risk of developing GESF compared to the latter (Siemens et al. 2018). This risk is associated with the APOL1 gene, with the G1 variant being present in 52% of patients with GESF (and in 18% of African Americans without GESF) and G2 being present in 23% and 15% of cases with and without GESF, respectively (Genovese et al. 2010b). The association of the two G1/G2 genetic variants confers a significant risk of GESF (more than a tenfold increase) in relation to carriers of none or only one genetic variant, which leads to an approximately onefold higher level of (Genovese et al. 2010b).

These findings were replicated in other studies featuring larger samples, such as the one reported by Kopp and colleagues (2011), in which APOL1 variants were identified in 271 African Americans, 168 Euro-Americans and 939 individuals in the control group. The presence of the G1/G2 variants in African Americans indicated a 17% higher risk for GESF and 29% higher risk for GESF associated with HIV infection. In addition, earlier age of disease onset was also confirmed, with 70% of the group being between 15–39 years old and exhibiting faster progression to ESRD, i.e. an average of 5 years for carriers of the G1/G2 variants and 13 years for those with no genetic variant.

6 APOL1, kidney disease and blacks

The presence of the G1 and G2 risky genetic variants are a protective factor against trypanosomiasis, but they entail a higher risk of different types of kidney disease in individuals of African ancestry, especially in the context of CKD associated with hypertension (SAH), glomerular diseases such as GESF, HIV nephropathy (HIVAN), and lupus nephritis associated with systemic lupus erythematosus in non-diabetic individuals (Siemens et al. 2018; Kruzel-Davilla et al. 2016).

In addition to CKD and ESRD, studies have also associated the presence of these genetic variants in the black population with susceptibility to systemic arterial hypertension, lupus nephritis, and poorer outcomes for kidney transplant recipients from donors of African descent, as they induce cytotoxicity and progressive kidney damage (Siemens et al. 2018; Freedman et al. 2018; Freedman et al. 2021). Alongside this increased risk, African descendants carrying the risky genetic variants manifest the disease at an earlier age, progress more rapidly to ESRD, and exhibit decreased renal survival in the long term (Kruzel-Davila et al. 2016; Freedman et al. 2021).

The reviewed studies predominantly report on the biological issues associated with APOL1 kidney damage and do not identify racism as a triggering factor associated with the faster progression of CKD. In the review, we found only one study (Norris et al. 2017) that highlights racism as a determinant of the progression of CKD in African Americans.

7 Final remarks

In clinical practice, it is of the utmost relevance to determine the degree of risk for conditions associated with a high prevalence for the development of severe renal disease in the population of African descendants, such as GESF, HIV and HIVAN nephropathy, CKD and ESRD associated with SAH, lupus nephritis, and poorer outcomes for renal transplant recipients. Based on recent data drawn from the literature, genetic investigation of APOL1, G1 and G2 in African descendants in Brazil is fundamental to the task of identifying the genetic risk of developing CKD exhibited by this population. The correct interpretation of the genetic analysis of APOL1 becomes a useful tool for risk prediction concerning different nephropathies and is critical with respect to the adoption of strategies for clinical management, the prevention of progression and risk reduction in African descendants as well as in the context of ensuring adequate planning and broad public health support for this high-risk population.

It should be noted that studies focusing on the black population remain incipient in the field of health in Brazil. This has a direct impact on public health policies and consequently on the planning of actions to prevent more advanced stages of CKD. Still, when analyzing the particularities of racism in Brazil, due to the precarious and unequal access of the black population to goods and services offered by the state, it is possible to make inferences regarding the high rates of morbidity and mortality exhibited by this population.

According to data drawn from the National Policy for Integral Health of the Black Population (2017), 12% to 14% of all deaths in Brazil are related to SAH, and, in general, this condition tends to be more complicated in blacks of both genders, while diabetes mellitus (type II) affects black men more frequently (9% more than white men).

In summary, in this sense, it is fundamental to emphasize the value of considering population-related aspects in the context of health research and studies.

Studies related to this topic demonstrate the need for specific public actions to be taken for certain populations, especially populations that have been marginalized by secular processes of exclusion and subjugation.

References

Arouet, François-Marie [Voltaire] (1773): *Philosophical Letters*. https://oll.libertyfund.org/title/fleming-the-works-of-voltaire-vol-xix-philosophical-letters, visited on 19 January 2022.

Alves-Silva, Juliana; Santos, Magda; Guimarães, Pedro; Ferreira, Alessandro; Bandelt, Hans-Jürgen; Pena, Sérgio; Prado, Vania (2000): "The ancestry of Brazilian mtDNA lineages". In: *American Journal of Human Genetics* 67. N. 2, pp. 444–61.

Brazil. (1988): *Constituição da República Federativa do Brasil de 1988* http://www.planalto.gov.br/ccivil_03/constituicao/constituicao.htm, visited on 22 February 2022.

Brazil. (2003): Law nº 10.678, 23 May 2003. http://www.planalto.gov.br/ccivil_03/leis/2003/l10.678.htm, visited on 22 February 2022.

Brazil. (2022): Law nº 28, 29 March 1884. https://www.al.sp.gov.br/norma/138408, visited on 22 February 2022.

Brazil. (2016): *Painel de Indicadores do SUS Nº10: Temático da Saúde da População Negra 7*, n. 10. Brasília: Ministério da Saúde.

Brazil. (2009): Decree nº 992, 13 May 2009. https://bvsms.saude.gov.br/bvs/saudelegis/gm/2009/prt0992_13_05_2009.html, visited on 22 February 2022.

Brazil. (2017): *Política Nacional de Saúde Integral da População Negra*". https://bvsms.saude.gov.br/bvs/publicacoes/politica_nacional_saude_populacao_negra_3 d.pdf, visited on 24 February 2022.

Brazil. (2021): *Sistema Único de Saúde (SUS): estrutura, princípios e como funciona*. https://www.gov.br/saude/pt-br/assuntos/saude-de-a-a-z/s/sus-estrutura-principios-e-como-funciona, visited on 22 February 2022.

Carneiro, Sueli. (2011): "Racismo, sexismo e desigualdade no Brasil". São Paulo: Selo Negro.

De Moura, Ronald; Coelho, Antônio; Balbino, Valdir; Crovella, Sergio; Brandão, Lucas (2015): "Meta-analysis of Brazilian Genetic Admixture and Comparison with other Latin America Countries". In: *American Journal of Human Biology* 27. Pp. 674–680.

Eknoyan, Garabed. (2013): "Kidney Diseases Improving Global Outcomes (KDIGO) 2012 Clinical Practice Guideline for the Evaluation and Management of Chronic Kidney Disease. From: Chronic kidney disease definition and classification: no need for a rush to judgment". In: *Kidney International Supplements* 3. N. 1, pp. S1–163.

Franco, Maria Helena; Brennan, Stephen; Chua, Kee; Kragh-Hansen, Ulrich; Callegari-Jacques, Sídia; Bezerra, Maria Zeneide; Salzano, Francisco (1999): "Albumin genetic variability in South America: Population distribution and molecular studies". In: *American Journal of Human Biology* 11. N. 3, pp. 359–66.

Freedman, Barry; Limou, Sophie; Ma, Lijun; Kopp, Jeffrey (2018): "APOL1-Associated Nephropathy: A Key Contributor to Racial Disparities in CKD". In: *American Journal of Kidney Diseases* 72. N. 5, pp. S8–S16.

Freedman, Barry; Kopp, Jeffrey; Sampson, Matthew; Susztak, Katalin (2021): "APOL1 at 10 years: progress and next steps". In: *Kidney International* 99. N. 6, pp. 1296–1302.

Friedman, David; Pollak, Martin (2011a): "Genetics of kidney failure and the evolving story of APOL1". In: *Journal of Clinical Investigation* 121. N. 9, pp. 3367–3374.

Friedman, David; Kozlitina, Julia; Genovese, Giulio; Jog, Prachi; Pollak, Martin (2011b). "Population-based risk assessment of APOL1 on renal disease". In: *Journal of the American Society of Nephrology* 22. N. 11, pp. 2098–2105.

Friedman, David; Pollak, Martin (2016): "Apolipoprotein L1 and Kidney Disease in African Americans." In: *Trends in Endocrinology & Metabolism* 27. N. 4, pp. 204–215.

Genovese, Giulio; Friedman, David; Ross, Michael; Lecordier, Laurence; Uzureau, Pierrick; Freedman, Barry; Bowden, Donald; Langefeld, Carl; Oleksyk, Taras; Knob, Andrea; Bernhardy, Andrea; Hicks. Pamela; Nelson, George; Vanhollebeke, Benoit; Winkler, Cheryl; Kopp, Jeffrey; Pays, Etienne; Pollak, Martin (2010a): "Association of Trypanolytic ApoL1 variants with kidney disease in African Americans". In: *Science* 329 N. 5993, pp. 841–845.

Genovese, Giulio; Tonna, Stephen; Knob, Andrea; Appel, Gerald; Katz, Avi; Bernhardy, Andrea; Needham, Alexander; Lazarus, Ross; Pollak, Martin. (2010b): "A risk allele for focal segmental glomerulosclerosis in African Americans is located within a region containing APOL1 and MYH9". In: *Kidney International* 78. N. 7, pp. 698–704.

Gomes, Laurentino. (2019): *Escravidão – Volume 1: Do primeiro leilão de cativo sem Portugal até a morte de Zumbi dos Palmares*. Brasilia: Globo Livros.

Gonçalves,Vanessa Faria; Stenderup, Jesper; Rodrigues-Carvalho, Cláudia; Silva, Hilton; Gonçalves-Dornelas, Higgor; Líryo, Andersen; Kivisild, Toomas; Malaspinas, Anna-Sapfo; Campos, Paula; Rasmussen, Morten; Willerslev, Eske; Pena, Danilo J Sergio. (2013): "Identification of Polynesian mtDNA haplogroups in remains of Botocudo Amerindians from Brazil". In: *Proceedings of the National Academy of Sciences* 110. N. 16, pp. 6465–6469.

Instituto Brasileiro de Geografia e Estatística [IBGE]. (2019): "Desigualdades Sociais por Raça no Brasil". https://biblioteca.ibge.gov.br/visualizacao/livros/liv101681_informativo.pdf, visited on 24 February 2022.

Instituto Brasileiro de Estatística e Geografia [IBGE]. (2022): "Projeção da população do Brasil e das Unidades da Federação". https://www.ibge.gov.br/apps/populacao/projecao//, visited on 22 February 2022.

Instituto Brasileiro de Estatística e Geografia [IBGE]. (2022): "Divisão Regional do Brasil". https://www.ibge.gov.br/geociencias/organizacao-do-territorio/divisao-regional/15778-di visoes-regionais-do-brasil.html, visited on 22 February 2022.

Johansen, Kristen; Chertow, Glenn; Foley, Robert; Gilbertson, David; Herzog, Charles; Ishani, Areef; Israni, Ajay; Ku, Elaine; Tamura, Manjula; Li, Shuling; Li, Suying; Liu, Jiannong; Obrador, Gregorio; O'Hare, Ann; Peng, Yi; Powe, Neil; Roetker, Nicholas; Peter, Wendy; Abbott, Kevin; Chan, Kevin; Schulman, Ivonne; Snyder, Jon; Solid, Craig; Weinhandl, Eric; Winkelmayer, Wolfgang; Wetmore, James (2021): "US Renal Data System 2020 Annual Data Report: Epidemiology of Kidney Disease in the United States". In: *American Journal of Kidney Diseases* 77. N. 4, pp. S1–S597.

Kehdy, Fernanda; Gouveia, Mateus; Machado, Moara; Magalhães, Wagner; Horimoto, Andrea; Horta, Bernardo; Moreira, Rennan; Leal, Thiago; Scliar, Marilia; Soares-Souza, Giordano; Rodrigues-Soares, Fernanda; Araújo, Gilderlanio; Zamudio, Roxana; Sant Anna, Hanaisa; Santos, Hadassa; Duarte, Nubia; Fiaccone, Rosemeire; Figueiredo, Camila; Silva; Thiago; Costa, Gustavo; Beleza, Sandra; Berg, Douglas; Cabrera, Lilia; Debortoli, Guilherme;

Duarte, Denise; Ghirtto, Silvia; Gilman, Robert; Gonçalves, Vanessa; Marrero, Andrea; Muniz, Yara; Weissensteiner, Hansi; Yeager, Meredith; Rodrigues, Laura; Barreto, Maurício; Lima-Costa, Fernanda; Pereira, Alexandre; Rodrigues, Maíra; Tarazona-Santos, Eduardo; Brazilian EPIGEN Project Consortium (2015): "Origin and dynamics of admixture in Brazilians and its effect on the pattern of deleterious mutations". In: *Proceedings of the National Academy of Sciences of the United States of America* 112. N. 28, pp. 8696–701.

Kopp, Jeffrey; Nelson, George; Sampath, Karmini; Johnson, Randall; Genovese, Giulio; An, Ping; Friedman, David; Briggs, William; Dart, Richard; Korbet, Stephen; Mokrzyuchi, Michele; Kimmel, Paul; Limou, Sophie; Ahuja, Tejinder; Berns, Jeffrey; Fryc, Justyna; Simon, Eric; Smith, Micahel; Trachtamn, Howard; Michel, Donna; Schelling, Jeffrey; Vlahov, David; Pollak, Martin; Winkler, Cheryl (2011): "APOL1 genetic variants in focal segmental glomerulosclerosis and HIV-associated nephropathy". In: *Journal of the American Society of Nephrology* 22. N. 11, pp. 2129–2137.

Kruzel-Davila, Etty; Wasser, Walter; Aviram, Sharon; Skorecki, Karl (2016). "APOL1 nephropathy: from gene to mechanisms of kidney injury". In: *Nephrology Dialysis Transplantation* 31. N. 3, pp. 349–358.

Moura, Lenildo; Prestes, Isaías; Duncan, Bruce; Thomé, Fernando; Schmidt, Maria Inês (2014): "Dialysis for end stage renal disease financed through the Brazilian National Health System, 2000 to 2012". In: *BMC Nephrology* 15. Pp. 111–116.

Munanga, Kabengele. (2009): "Origens africanas do Brasil contemporâneo: histórias, línguas, culturas e civilizações". São Paulo: Global Editora.

Munanga, Kabengele. (2004): "Rediscutindo a mestiçagem no Brasil: identidade nacional versus identidade negra". Belo Horizonte: Autêntica Editora.

Naslavsky, Michel; Yamamoto, Guilherme; de Almeida, Tatiana; Ezquina, Suzana; Sunaga, Daniele; Pho, Nam; Bozoklian, Daniel; Sandberg, Tatiana; Brito, Luciano; Lazar, Monizel Bernardo, Danilo; Amaro Jr. Edson; Duarte, Yeda; Lebrão, Maria Lúcia; Passos-Bueno, Maria Rita; Zatz, Mayana (2017): "Exomic variants of an elderly cohort of Brazilians in the ABraOM database". In: *Human Mutation* 38. N. 7, pp. 751–63.

Neves, Precil; Sesso, Ricardo; Thomé, Fernando; Lugon, Jocemir; Nascimento, Marcelo (2021): "Brazilian dialysis survey 2019". In: *Brazilian Journal of Nephrology* 43. N. 2, pp. 217–227.

Norris, Keith; Williams, Sandra; Rhee, Connie; Nicholas, Susanne; Kovesdy, Csaba; Kalantar-Zadeh, Kamyar; Boulware, Ebony (2017): "Hemodialysis disparities in African Americans: the deeply integrated concept of race in the social fabric of our society". In: *Seminars in Dialysis*. N. 3, pp. 213–223.

Parsa, Afshin; Kao, Linda; Xie, Dawei; Astor, Brad; Li, Man; Hsu, Chi-yuan; Feldman, Harold; Parekh, John; Greene, Tom; Fink, Jeffrey; Anderson, Amanda; Choi, Michael; Wright Jr, Jackson; Lash, James; Freedman, Barry; Ojo, Akinlolu; Winkler, Cheryl; Raj, Dominic; Kopp, Jeffrey; He, Jiang; Jensvold, Nancy; Tao, Kaixiang; Lipkowitz, Michael; Appel, Lawrence; AASK Study Investigators; CRIC Study Investigators (2013): "APOL1 risk variants, race, and progression of chronic kidney disease". In: *New England Journal of Medicine* 369. N. 23, pp. 2183–2196.

Pena, Sérgio. (2008): "Humanidade sem raças?". São Paulo: Editora Publifolha.

Riella, Cristian; Siemens, Tobias; Wang, Minxian; Campos, Rodrigo; de Moraes, Thyago; Riella, Leonardo; Friedman, David; Riella, Miguel; Pollak, Martin (2019):

"APOL1-Associated Kidney Disease in Brazil". In: *Kidney International Reports* 4. N. 7, pp. 923–929.

Ruiz-Linares, Andrés; Adhikari, Kaustubh; Acunã-Alonzo, Victor; Quinto-Sanchez, Mirsha; Jaramillo, Claudia; Arias, William; Fuentes; Macarena; Pizarro, María; Everardo, Paola; de Avila, Francisco; Gómez-Valdés, Jorge; León-Mimila, Paola; Hunemeier, Tábita; Ramallo, Virginia; de Cerqueria, Caio; Burley, Mari-Wyn; Konca, Esra; de Oliveira, Marcelo; Veronez, Mauricio; Rubio-Codina, Marta; Attanasio, Orazio; Gibbon, Sahra; Ray, Nicolas; Gallo, Carla; Poletti, Giovanni; Rosique, Javier; Schuler-Faccini, Lavinia; Salzano, Francisco; Bortolini, Maria-Cátira; Canizales-Quinteros, Samuel; Rothhammer, Francisco; Bedoya, Gabriel; Balding, David, Gonzales-José, Rolando (2014). "Admixture in Latin America: Geographic Structure, Phenotypic Diversity and Self-Perception of Ancestry Based on 7,342 Individuals". In: *PLOS Genetics* 10. N. 9, e1004572.

Saran, Rajiv; Li, Yi; Robinson, Bruce; Abbott, Kevin; Agodoa, Lawrence; Ayanian, John; Bragg-Gresham, Jennifer; Balkrishnan, Rajesh; Chen, Joline; Cope, Elizabeth; Eggers, Paul; Gillen, Daniel; Gipson, Debbie; Hailpern, Susan; Hall, Yoshio; He, Kevin; Herman, William; Heung, Michael; Hirth, Richard; Hutton, David; Jacobsen, Steven; Kalantar-Zadeh, Kamyar; Kovesdy, Casba; Lu, Yee; Molnar, Miklos, Morgenstern, Hal; Nallamouthu, Brahmajee; Nguyen, Danh; O'Hare, Ann; Plattner, Brett; Pisoni, Ronald; Port, Friedrich; Rao, Panduranga; Rhee, Connie; Sakhuja, Ankit; Schaubel, Douglas; Selewski, David; Shahinian, Vahakn; Sim, John; Song, Peter; Streja, Elani; Tamura, Manjula; Tentori, Francesca; White, Sarah; Woodside, Kenneth; Hirth, Richard (2016): "US Renal Data System 2015 Annual Data Report: epidemiology of kidney disease in the United States". In: *American Journal of Kidney Diseases* 67. N. 3, pp. S1-S434.

Siemens, Tobias; Riella, Miguel; de Moraes, Thyago; Riella, Cristian (2018): "APOL1 risk variants and kidney disease: what we know so far". In: *Brazilian Journal of Nephrology* 40. N. 4, pp. 388–402.

Sociedade Brasileira de Nefrologia [SBN]. (2022): "Sociedade Brasileira de Nefrologia – história". https://www.sbn.org.br/a-sbn/quem-somos/, visited on: 22 February 2022.

Thornton, John K. (2010): "Angola e as origens de Palmares". In: Gomes, Flávio (ed), *Mocambos de Palmares. Histórias e fontes* (séculos XVI-XIX). Rio de Janeiro: 7 Letras.

Nchangwi Syntia Munung
Ethnicity, Geographical Region and Ancestry as Population Level Descriptors for Genomics Studies in Africa: Public Engagement is Needed

Abstract: In designing and reporting population genetic studies, researchers use population descriptors such as race, ethnicity, geographical location, and ancestry/biogeographical ancestry as analytical variables. These population level descriptor(s) are complex social constructs and there are ongoing conversations concerning their use in population genomics. As academics debate and seek consensus regarding the population level descriptors that can be best used to account for genetic variation both within and between population groups, they must reflect on whether these descriptors are socially and culturally acceptable for studying communities. This would require science outreach activities to consult with different stakeholders regarding what each population descriptor means to them. These outreach activities should engage the public in understanding how and why each descriptor is used for certain studies. This is particularly the case in situations in which 1) a descriptor may have pejorative undertones due to its historical background or 2) the study pertains to a disease that is stigmatized, so that the use of a particular descriptor may further perpetuate stigma and/or discrimination.

1 Introduction

Race, ethnicity, ancestry and continental labels such as "Africans", "Asians", and "Europeans" have consistently been used as bounded analytical variables in population genomics studies. Perhaps there are justifiable reasons for doing so. From a scientific and equity-based standpoint, these bounded variables are a guide to ensuring that diverse population groups are included in genomics studies and by extension benefit from medical advances in genetics. The increased use of these variables is also a response to concerns that genomics is failing in terms of diversity and that many genome wide association studies (GWAS) have mainly focused on populations of European descent (Gurdasani

Nchangwi Syntia Munung, University of Cape Town

https://doi.org/10.1515/9783110765120-007

et al. 2019; Popejoy and Fullerton 2016). It has therefore been argued that limited representation of populations from Africa, Asia, The Americas, and Oceania in genomic studies could significantly limit our understanding of the genetic determinants of disease as well as create a global genomics divide (Singer and Daar 2001; Sirugo et al. 2019). All these factors have cumulated in a major drive to conduct genetic studies in certain populations, defined in terms of geographical location, ethnicity, race, tribal community, or ancestry. For example, the Human Heredity and Health in Africa (H3Africa) initiative is a "continental level" study with a focus on genomics and environmental determinants of common diseases in African populations (H3Africa Consortium 2014), while the African Genome variation project is designed and reported in terms of ethno-linguistic groups (Gurdasani et al. 2015).

Generally, social scientists and anthropologists describe race and ethnicity as bounded and/or as social constructs used to categorize and characterize seemingly distinct populations (Ifekwunigwe et al. 2017). In the case of race, this classification is often based on phenotypic features like skin color and, to some extent, hair texture and eye color. Ethnicity, on the other hand highlights distinct cultural expressions. The use of race as a proxy in genomics research has been strongly criticized by biologists, social scientists, bioethicists and anthropologists (Bonham et al. 2018; Foster and Sharp 2004; Mohsen 2020) because race is considered a social and political construct and therefore should not be used to define genetic origins.

Considering the existence of an extensive debate concerning the use of race as a proxy or analytical variable in genomics (Bliss 2020; Maglo et al. 2016; Sirugo and Wonkam 2022), it may be better to shift attention from race to other emerging or preferred proxies, such as ethnicity, ancestry and geographical region. In this chapter, I reflect on ethnicity and ancestry as two proxies that, in combination with geographical location, are used extensively in reporting the outcomes of genomics studies conducted in Africa. For example, some studies involving populations from different African countries have reported outcomes using ethnicity (language family), geographical location (country) admixed populations and a combination of ethnicity and geographical location, such as by referring to Bantu-speaking individuals from Uganda, people from Cote d'Ivoire, the Central-West African group, and non-Niger-Congo (Campbell and Tishkoff 2008; Choudhury et al. 2020).

2 Ethnicity as a proxy or analytical variable in genomics research in Africa

Debates concerning whether ethnicity is an ideal proxy for population genetic studies and persons remains ongoing (Kwan 2020; Manica et al. 2005). The concerns that have been expressed thus far include the possibility of stigmatization of certain ethnic groups (de Vries et al. 2012), the potential of perpetuating a misguided notion that discrete genetic groups exist (Bonham et al. 2018), that ethnicity, like race, is a socio-political construct and in some instances a result of interventions by colonial authorities (Yéré et al. 2022). Ethnicity is a multidimensional construct that mainly refers to a defined group of people who share a common and distinctive cultural attribute, such as a language, belief system, or set of customs and traditions. Therefore, people from the same ethnic group may, or may not, have a common genetic origin. For example, I am Bantu not because of my genetic makeup but my cultural heritage, although admittedly there is some degree of genetic similarity among the Bantus, since many members of this group share a common ancestor.

If ethnicity is a social and political construct, what is its nature when used as an analytical category in population genetic studies? In addition, if we agree that ethnicity should not be used as an analytical category in population genetics, what options are available for defining populations, especially in terms of understanding genetic variation and their role in health and disease? These questions will undoubtedly benefit from a multidisciplinary conversation between biomedical scientists, social scientists and historians. What will be important in such conversations is to emphasize the end goal of population genomics studies and if it is possible to reach that end goal without reference to ethnicity.

Less than two decades ago, there was growing concern that population genomics studies have mainly focused on people of European descent and that important genetic associations and variations that are present in other populations may have been missed. This has implications for equitable access to genomics/precision medicine. For example, while it was widely believed that cystic fibrosis was rare in non-European populations, and so GWAS investigating cystic fibrosis manly focused on populations of European descent, recent studies have instead shown that unique pathogenic variants of the CFTR gene are found in African populations (Mutesa and Bours 2009; Owusu et al. 2020; Stewart and Pepper 2016). The outcomes of these studies have implications for the diagnosis of cystic fibrosis in persons of African descent as well as for genetic counselling. For non-monogenetic conditions such as diabetes, it is difficult to see how this approach may apply, as the disease causation may be due to genetic factors and the social determinants of health. Therefore, using ethnicity as a proxy may require researchers to define the ethnic groups included in their studies. This necessity is due to the fact that there are thousands of ethnic groups in Africa, and while a group of people may belong to one ethnolinguistic population (e. g., Bantu expression), they may also belong to different ethnic groups, each of which has different social values, behavioral patterns and lifestyles.

As an analytical variable, the use of ethnicity in reporting genomics studies can help inform public health policy regarding genomics; nonetheless, this can only be effective if studies are clear regarding their definitions of ethnic groups. There are thousands of ethnic groups in Africa and although these groups have been categorized as Afroasiatic, Khoisan, Niger-Congo and Nilo-Saharan, each of these four ethno-linguistic groups is composed of several ethnicities (Goodman 1970; Greenberg 1948). For example, the Bantus, a commonly used proxy in African population genetic studies, belong to the Niger-Congo ethno-linguistic category and consist of approximately four hundred ethnicities, each with defined cultural attributes. There are therefore likely to be differences in disease onset and progression both within and between ethnic groups based on specific anthropometric phenotypes as well as social, behavioral, and environmental factors, even if studies indicate that the larger ethnic category (for example, Bantu) has a high genetic predisposition to developing a particular disease condition, such as diabetes or hypertension.

One of the major challenges associated with using ethnicity as a proxy or analytical category in population genomics is that some ethnic names have historical undertones that may be hurtful to some persons who identify with those ethnic groups. For example, in the 20th century, race and ethnicity were used during the apartheid era as tools for promoting segregation and discrimination, and the word Bantu (which means people in many Bantu languages) was used by the apartheid government as an ethno-racial category for black South Africans. As a result, some apartheid liberation movements opted to use "Africans" to describe racially oppressed groups, or "blacks" to refer to Bantu-speaking South Africans. In modern day South Africa, the terms "African" or "black African" are used in official documents as a racial/ethnic category for "African or black" South Africans. What remains unclear is whether black South Africans are concerned by the use of Bantu (or Bantu expression) as a proxy in genomics and health research. In other parts of Africa, "Bantu" may have a positive connotation or a neutral undertone. For example, the Somali Bantus, a marginalized minority group and referred to by some Somalis as *sheegato* or *sheegad* (pretenders) because they are not ethnically Somalians or Afroasiatic (Webersik 2004), have continued to refer to themselves as Bantu. In some instances, they have sought resettlement in areas that they consider to be ancestral "Bantu" land.

One may therefore imagine that the use of the term Bantu to describe this population group in genetic studies would be acceptable to them.

3 Geographical regions as an analytical variable in genomics research in Africa

Geographical regions, such as continental labels, are an emerging analytical variable or proxy in population genomics research (Byeon et al. 2021). The "African genome" is constantly referenced as key to understanding human genetic diversity; there are also region-specific population genomics initiatives such as the "Southern African Human Genome Programme", which focuses specifically on "Southern African genomes" (Pepper 2011), and a case has been made for genetics studies to investigate Somali populations living in Somalia, although the authors acknowledged the existence of studies that included Somali populations in the diaspora (Ali et al. 2020). All these emphasize the notion of genomic sovereignty and create a sense of genomic groups as biopolitical entities that could be used as scientific, health-related and economic resources by nation states (Benjamin 2009; Séguin et al. 2008). For instance, in 2019, a research institution in the United Kingdom was accused of misusing African DNA after whistle blowers revealed that there were plans to commercialize a new gene chip developed from samples collected from some genomics studies conducted in African countries (Stokstad 2019). The concern was that samples were sent to the UK institution for research purposes only, and there was no legal agreement with the partner institutions in Africa to use the donated DNA samples to develop commercial products. It was also argued that the UK institution did not obtain consent from the hundreds of African people whose DNA had been use used to develop the gene chip. Accordingly, some African research institutions requested that the donated DNA samples be returned, as their use for commercial purposes had serious legal and ethical consequences. This raises question of whether nation-states, communities and institutions can make claims regarding the sharing of profits or other benefits derived from the use of DNA samples collected from a specific country or geographical region.

Recently, some individuals have questioned what certain geographical labels mean in the context of genomics research. For example, what does "African" in the phrases "African genome" or "African genetic database" signify (Yéré et al. 2022). Worthy of note is the fact that the use of geographical labels is not peculiar to Africa; there is also the "IndiGen" program, which aims to perform whole-genome sequencing of thousands of individuals representing diverse ethnic groups from India (Jain et al. 2021), as well as "The Asian Genomes" project (Tang 2020) and the "European genome-phenome archives" (Lappalainen et al. 2015). Arguments related to identity similar to that advanced by Yéré and colleagues (2022) may apply to some of these population groups. However, the bigger question, I think pertains to whether scientists are intentionally creating a biopolitical entity when they seek to understand human genetic variation.

4 Ancestry as a preferred proxy in population genomics research

A popular category and one to which many researchers seem to be warming up as an analytical variable is ancestry (Byeon et al. 2021; Popejoy et al. 2020). It is not uncommon to see genomics studies refer to people of African ancestry as a population group. As a proxy for genetics studies, ancestry allows scientists to examine genetic diversity or a predisposition toward certain phenotypes based on geographical ancestral origins. This has been the case for sickle cell disease, for example, which studies have established is more common in people of African descent (Esoh and Wonkam 2021). The use of bio-geographical ancestry offers a solution for studies involving sympatric populations as it reduces the risk of discrimination against historically disadvantaged groups or minority populations.

The use of ancestry as a population level descriptor for genomics studies is not without controversy or concerns. Firstly, ancestry has different connotations. It could be biological, historical, cultural, geographical, political or a combination of these dimensions. Therefore, it is not surprising that the term bio-geographical ancestry is increasingly gaining prominence in population genetics studies. Secondly, the origin of a particular genetic mutation does not necessarily where that mutation is most common today; therefore, assigning a particular genetic combination to a geographical area is problematic.

5 Public engagement: a way forward for identifying (in)appropriate population level descriptors in genomics

As we foreseeably enterin era of academic debates on appropriate population-level descriptors for genetic studies, all stakeholders, irrespective of their disciplinary membership or roles, must remember that each variable is likely to be a multiplex concept and that its use, application and even definition may vary depending on the research context. Our focus should be on how the use of each variable may eventually inform research translation in terms of policy and practice. The imprecise use of any population level descriptor has the potential to introduce significant limitations related to equitable global access to precision medicine (Bonham et al. 2018). Therefore, alongside the public or study communities, policy makers and researchers must reflect on the assumptions and limitations that each variable introduces to a specific study, including whether its use may lead to discrimination and stigma or promote global inequities in science and health. Public outreach programs to educate and engage with various stakeholders concerning the ways in which scientists use these variables are necessary. Scientists should be able to explain to their study populations how and why they have opted for one variable over another and the implications of these choices in terms of research translation.

6 Conclusion

Genomic research in diverse populations has led to numerous advances in our understanding of human biological diversity, health and disease. When designing and/or reporting population genetic studies, researchers tend to use population descriptors such as race, ethnicity, geographical location and ancestry. Each of these descriptors plays a role in understanding human genetic variation and its contribution to health and disease. Nonetheless, researchers should be conscious of the social and health justice implications of using different population level descriptors. One way of doing so is to through public engagement so as to gain an understanding of how populations engage the public regarding what each population descriptor means in the context of genomics and why it is used in certain studies. It would require engaging with the public to gain a finer understanding which variables may be most appropriate an socially/culturally acceptable . This is particularly the case in situations whereby 1) a descriptor may have pejorative undertones because of its historical background (for example, the use of the term Bantu during the apartheid era), 2) the study focuses on a disease condition that is already stigmatized, and so a particular descriptor may further perpetuate stigma and/or discrimination, and 3) when use of the descriptor may promote inequities in health among different population groups and create a genomic divide.

Acknowledgements

I am grateful to Kevin Esoh for lengthy discussions concerning the use of ethnicity in genomics research. I would like to acknowledge the discussions and exchanges of ideas that occurred at the 'Situating the African Genome' interdisciplinary workshop held at STIAS, South Africa, in March 2020, which as funded by Program Point Sud and the German Research Council.

I am supported by a DSI-Africa grant from the National Human Genome Research Institute and the Office of the Director of the National Institutes of Health (award number 1U01MH127692–01). The views expressed in this chapter are not necessarily the views of the funder. The funder had no role in the preparation of the manuscript or the decision to publish it.

References

Ali, Abshir A.; Mikko Aalto, Jon Jonasson & Abdimajid Osman. (2020): "Genome-wide analyses disclose the distinctive HLA architecture and the pharmacogenetic landscape of the Somali population". In: *Scientific Reports* 10. N. 1, 5652.

Ananyo, Choudhury; Aron, Shaun; Botigué, Laura; Sengupta, Dhriti; Botha, Gerrit; Bensellak, Taoufik; Wells, Gordon; Kumuthini, Judit; Shriner, Daniel; Fakim, Yasmina J.; Ghoorah, Anisah W.; Dareng, Eileen: Odia, Trust; Falola, Oluwadamilare; Adebiyi, Ezekiel; Hazelhurst, Scott; Mazandu, Gaston; Nyangiri, Oscar A.; Mbiyavanga, Mamana; Benkahla, Alia; Kassim, Samar K.; Mulder, Nicola; Adebamowo, Sally N.; Chimusa, Emile R.; Rotimi, Charles; Ramsay, Michèle; Adeyemo, Adebowale A.; Lombard, Zané; & Hanchard, Neil A. (2020): "High-depth African genomes inform human migration and health". In: *Nature* 586. N. 7831, pp. 741–748.

Benjamin, Ruha. (2009): "A Lab of Their Own: Genomic sovereignty as postcolonial science policy". In: *Policy and Society* 28. N. 4, pp. 341–355.

Bliss, Catherine. (2020): "Conceptualizing Race in the Genomic Age". In: *Hastings Center Report* 50. Pp. S15-S22.

Bonham, Vence L.; Green, Eric D.; & Pérez-Stable, Eliseo J. (2018): "Examining How Race, Ethnicity, and Ancestry Data Are Used in Biomedical Research". In: *JAMA* 320. N. 15, pp. 1533–1534.

Byeon, Yen Ji Julia; Islamaj, Rezarta; Yeganova, Lana; Wilbur, W. John; Zhiyong, Lu; Brody, Lawrence C.; & Bonham, Vence L. (2021): "Evolving use of ancestry, ethnicity, and race in genetics research—A survey spanning seven decades". In: *The American Journal of Human Genetics* 108. N. 12, pp. 2215–2223.

Campbell, Michael C.; & Tishkoff, Sarah A. (2008): "African genetic diversity: implications for human demographic history, modern human origins, and complex disease mapping". In: *Annu Rev Genomics Hum Genet* 9. Pp. 403–433.

de Vries, Jantina; Jallow, Muminatou; Williams, Thomas N.; Kwiatkowski, Dominic; Parker, Michael; & Fitzpatrick, Raymond. (2012): "Investigating the potential for ethnic group harm in collaborative genomics research in Africa: Is ethnic stigmatisation likely?". In: *Social Science & Medicine* 75. N. 8, pp. 1400–1407.

Esoh, Kevin; & Wonkam, Ambroise. (2021): "Evolutionary history of sickle-cell mutation: implications for global genetic medicine". In: *Hum Mol Genet* 30. N. R1, pp. 119-r128.

Foster, Morris W.; & Sharp, Richard. (2004): "Beyond race: towards a whole-genome perspective on human populations and genetic variation". In: *Nat Rev Genet* 5. N. 10, pp. 790–796.

Goodman, Morris. (1970): "Some Questions on the Classification of African Languages". In: *International Journal of American Linguistics* 36. N. 2, pp. 117–122.

Greenberg, Joseph H. (1948): "The Classification of African Languages". In: *American Anthropologist* 50. N. 1, pp. 24–30.

Gurdasani, Deepti; Barroso, Inês; Zeggini, Eleftheria; & Sandhu. Manjinder S. (2019): "Genomics of disease risk in globally diverse populations". In: *Nature Reviews Genetics* 20. N. 9, pp. 520–535.

Gurdasani, Deepti; Carstensen, Tommy; Tekola-Ayele, Fasil; Pagani, Luca; Tachmazidou, Ioanna; Hatzikotoulas, Konstantinos; Karthikeyan, Savita; Iles, Louise; Pollard, Martin O.; Choudhury, Ananyo; Ritchie, Graham R. S.; Xue, Yali; Asimit, Jennifer; Nsubuga, Rebecca N.; Young, Elizabeth H.; Pomilla, Cristina; Kivinen, Katja; Rockett, Kirk; Kamali, Anatoli; Doumatey, Ayo P.; Asiki, Gershim; Seeley, Janet; Sisay-Joof, Fatoumatta; Jallow, Muminatou; Tollman, Stephen; Mekonnen, Ephrem; Ekong, Rosemary; Oljira, Tamiru; Bradman, Neil; Bojang, Kalifa; Ramsay, Michele; Adeyemo, Adebowale; Bekele, Endashaw; Motala, Ayesha; Norris, Shane A.; Pirie, Fraser; Kaleebu, Pontiano; Kwiatkowski, Dominic; Tyler-Smith, Chris; Rotimi, Charles; Zeggini, Eleftheria; & Sandhu, Manjinder S. (2015): "The African Genome Variation Project shapes medical genetics in Africa". In: *Nature* 517. N. 7534, pp. 327–332.

H3Africa Consortium. (2014): "Enabling the genomic revolution in Africa". In: *Science* 344. N. 6190, pp. 1346–1348.

Ifekwunigwe, Jayne O., Wagner, Jennifer K.; Yu, Joon-Ho; Harrell, Tanya M.; Bamshad, Michael J.; & Royal, Charmaine D. (2017): "A Qualitative Analysis of How Anthropologists Interpret the Race Construct". In: *Am Anthropol* 119. N. 3, pp. 422–434.

Jain, Abhinav; Bhoyar, Rahul C.; Pandhare, Kavita; Mishra, Anushree; Sharma, Disha ; Imran, Mohamed; Senthivel, Vigneshwar; Kumar, Mohit; Divakar, Mercy Rophina; Bani Jolly; Arushi, Batra; Sumit Sharma; Sanjay, Siwach; Arun G., Jadhao; Nikhil V., Palande; Ganga, Nath Jha; Nishat, Ashrafi; Prashant, Kumar; Mishra, A. K. Vidhya;, Suman, Jain; Debasis, Dash; Nachimuthu, Senthil Kumar; Andrew, Vanlallawma; Ranjan, Jyoti Sarma; Lalchhandama, Chhakchhuak; Shantaraman, Kalyanaraman; Radha, Mahadevan; Sunitha, Kandasamy, Pabitha, B. M.; Raskin Erusan, Rajagopal; Ezhil, Ramya J.; Devi, P.; Bajaj, Nirmala; Anjali, Vishu Gupta; Mathew, Samatha; Goswami, Sangam; Mohit, Mangla; Savinitha Prakash, Kandarp; Joshi, Sreedevi S.; Meyakumla, Devarshi; Gajjar, Ronibala; Soraisham, Rohit; Yadav, Yumnam Silla Devi; Aayush, Gupta; Mitali, Mukerji; Sivaprakash Ramalingam; Binukumar, B. K.; Vinod, Scaria; & Sridhar, Sivasubbu. (2021): "IndiGenomes: a comprehensive resource of genetic variants from over 1000 Indian genomes". In: *Nucleic Acids Res* 49. N. D1, pp. D1225-D1232.

Kwan, Hoi Shan. (2020): "Genetics: Ethnicity". In Danan Gu & Matthew E. Dupre (eds.), *Encyclopedia of Gerontology and Population Aging*, 1–6. Cham: Springer International Publishing.

Lappalainen, Ilkka; Jeff, Almeida-King; Vasudev, Kumanduri; et al. (2015): "The European Genome-phenome Archive of human data consented for biomedical research". In: *Nat Genet* 47. N. 7, pp. 692–695.

Maglo, Koffi N., Mersha, Tesfaye B.; & Martin, Lisa J. (2016): "Population Genomics and the Statistical Values of Race: An Interdisciplinary Perspective on the Biological Classification of Human Populations and Implications for Clinical Genetic Epidemiological Research". In: *Frontiers in Genetics* 7. N. 22, pp. 1-13, doi: 10.3389/fgene.2016.00022

Manica, Andrea, Prugnolle, Franck, & Balloux, François. (2005): "Geography is a better determinant of human genetic differentiation than ethnicity". In: *Hum Genet* 118. N. 3, pp. 366–371.

Mohsen, Hussein. (2020): "Race and Genetics: Somber History, Troubled Present". In: *Yale J Biol Med* 93. N. 1, pp. 215–219.

Mutesa, Léon; & Bours, Vincent. (2009): "Diagnostic Challenges of Cystic Fibrosis in Patients of African Origin". In: *Journal of Tropical Pediatrics* 55. N. 5, pp. 281–286.

Owusu, Sandra Kwarteng; Morrow, Brenda M.; White, Debbie; Klugman, Susan; Vanker, Aneesa; Gray, Diane; & Zampoli, Marco. (2020): "Cystic fibrosis in black African children in South Africa: a case control study". In: *Journal of Cystic Fibrosis* 19. N. 4, pp. 540–545.

Pepper, Michael Sean. (2011): "Launch of the Southern African Human Genome Programme". In: *S Afr Med J* 101. N. 5, pp. 287–288.

Popejoy, Alice B.; & Fullerton, Stephanie M. (2016): "Genomics is failing on diversity". In: *Nature* 538. N. 7624, pp. 161–164.

Séguin, Béatrice; Hardy, Billie-Jo; Singer; Peter A.; & Daar, Abdallah S. (2008): "Genomic medicine and developing countries: creating a room of their own". In: *Nat Rev Genet* 9. N. 6, pp. 487–493.

Singer, Peter A.; & Daar, Abdallah S. (2001): "Harnessing Genomics and Biotechnology to Improve Global Health Equity". In: *Science* 294. N. 5540, pp. 87–89.

Sirugo, Giorgio; Scott M. Williams; & Tishkoff, Sarah A. (2019): "The Missing Diversity in Human Genetic Studies". In: *Cell* 177. N. 1, pp. 26–31.

Sirugo, Giorgio; & Wonkam, Ambroise. (2022): "Beyond Race: A Wake-up Call for Drug Therapy Informed by Genotyping". In: *Annals of Internal Medicine*, ahead of print, doi: 10.7326/M22–1827.

Stewart, Cheryl; & Pepper, Michael S. (2016): "Cystic fibrosis on the African continent". In: *Genetics in Medicine* 18. N. 7, pp. 653–662.

Stokstad, Erik. (2019): "Genetics lab accused of misusing African DNA". In: *Science* 366. N. 6465, pp. 555–556.

Tang, Lin. (2020): "Asian genomes". In: *Nature Methods* 17. N. 2, pp. 130–130.

Webersik, Christian. (2004): "Differences that Matter: the Struggle of the Marginalised in Somalia". In: *Africa* 74. N. 4, pp. 516–533.

Yéré, Henri-Michel, Machirori, Mavis; & de Vries, Jantina. (2022): "Unpacking race and ethnicity in African genomics research". In: *Nat Rev Genet*. Ahead of print, https://doi.org/10.1038/s41576-022-00506-4.

Katya Gibel Mevorach
Racism, Pedagogy and Willful Ignorance

Abstract: This chapter argues for intervening against and discrediting the race concept in science education and research. The histories, contexts and consequences of racism evidence the necessity of vigilance in teaching and uprooting false ideas which otherwise serve as the premise and departure point for prescriptive questions and interpretation. This means candidly confronting dismantling racism and making visible its trace. The question is always, what difference does difference make?

1 Introduction

The confusion of the relationship between the "noun" race – a classification system based on ancestry – with the "practice" of "racism", is reinforced daily by lazy language and willful ignorance. Some of this can be traced to professional stakeholders whose entire careers cling to the edifice promoting the significance of race as "identity" and "identifier" rather than the imposition or adaptation of race as an "identification" and therefore a "label". The cottage industry around "race" and corollary adjectives generated by the concept "race" offer misleading explanatory reasons for discrimination and disparity. Despite the proliferation of texts and talks discrediting the race concept, including books, peer-reviewed academic articles, conference presentations and podcasts highlighting the consequences of racism, the financial investment in cultivating arguments that acknowledge "race" is a "social construction", a signifier and nevertheless proceed to treat the monstrous concept as an inherent feature of human lives continues to generate profit. One might speculate on a comparison of royalty revenues between Robin Di Angelo's *White Fragility* (Di Angelo 2018) and Law Professor Dorothy Roberts *Fatal Invention* (Roberts 2012). Both books are reader friendly and accessible to a non-academic audience but only one author is a media sensation – parenthetically it is fascinating to consider the racializing dynamics that give broad visibility to Di Angelo, a white woman marketing a primordial argument about whiteness and more discrete attention to Roberts, a Black woman. I imagine a counter-argument that one has a better agent.

The reductionist ahistorical viewpoint is that the race concept is timeless and the producer, rather than the product, of racism. This perspective puts the

Katya Gibel Mevorach, Grinnell College

https://doi.org/10.1515/9783110765120-008

idea of race outside of history and political economy and overshadows the insidious evil of racism as an active force. To write this short essay, which is adapted from a presentation, I have to preface with the caveat from psychiatrist and political theorist Franz Fanon, who opens *Black Skin, White Masks* with the comment that "[t]here are too many idiots in this world. And having said it, I have the burden of proving it"[1] (Fanon 1952).

2 The challenge: ideological shibboleths

Current debates over the relationship of class and race seem designed to manipulate and exacerbate ideological shibboleths. This essay draws on a co-taught course which is cross listed between American Studies, Anthropology, and Biology. In fact, as we designed our curricular project against race-thinking, we understood from the outset that its symbolic leverage was the umbrella of Biology. Therefore while *Introduction to Biology* is a rarely waived prerequisite, leniency is granted toward a social science prerequisite because this does not guarantee that students acquired insights into the need for historicization and context; on the contrary, more and more students are attracted to a pedagogy promoting advocacy rather than accuracy. In this essay, my goal is to underline several key points which may be useful for educators and graduate students shaping an approach to teaching, studying or participating in biomedical research and health professions that engages, rather than erases, issues of racism. Note immediately – this has nothing to do with personal behavior or subjective identities. It is explicitly recognizing that everything to do with biology and genetics has to do with identifying, comparing and differentiating bodies. Scientists and the fields of science are therefore not immune to, or innocent from, the socio-economic, political and cultural worlds in which intellectual and physical work is carried out.[2]

The colleague with whom I teach is a molecular biologist who attended an American Studies faculty reading workshop almost twenty years ago which fo-

---

1 Fanon first asks: "Why write this book? No one has asked me for it. Especially those to whom it is directed.
Well?" (Fanon 1977). The paragraph in the original French reads as follows: "Pourquoi écrire cet ouvrage ? Personne ne m'en a prié. Surtout pas ceux à qui il s'adresse. Alors ? Alors, calmement, je réponds qu'il y a trop d'imbéciles sur cette terre. Et puisque je le dis, il s'agit de le prouver" (Fanon 1952, p. 5).
2 K.Gibel Mevorach (2007) Reflections on 'Race' and the Biologization of Difference." In Race and Contemporary Medicine. Ed. Sander Gilman, London: Routledge.

cused on the Africanist presence in shaping ideas about American identities, literature, history and economics.[3] In the context of slavery – that is human trafficking and unpaid forced labor – and post-Civil War ambition of white Anglo Protestants to maintain political and economic power, it was important to understand the centrality of racism and the imperative power of both legislative policies and authority of scientists' ideas about human difference. An important goal of the workshop, therefore, was to underline the importance of attending to the inseparable histories of intersection between Anglo-Americans and Afro-Americans – insidious, complex, messy, yes, but nevertheless a relationship aptly described by Barbara Fields and Karen Fields, "lives lived separately together" (Fields and Fields 2012).

A few years later, my colleague invited me to craft a course together which would take up threads from a workshop about which, frankly, I had filed and shelved. This was due in part because of a tediousness to carry the same banner encouraging the crossing – trespassing actually – disciplinary boundaries and actively arguing for the study of race as a consequence of racism. The first step in deciding to take on creating a course is to determine shared objectives and a common purpose. The second step, departmental and administrative approval, was easier at our liberal arts college because of a push for inter-departmental projects and a President whose credentials impressively included MD, MBA and PhD[4]. Raynard Kington had been recruited from a senior position at the National Institute of Health so his background in administration and health would make him an attractive guest to visit with our students each semester. The point is that purposeful determination and confidence, combined with seeking and receiving support, is an effective way to percolating ideas, even when the effort sometimes feels Sisyphean.

We begin by acknowledging our intimidation with each other's academic disciplines. The sentiment has not changed over time and turns out to be an asset: most of our students are interested in being science majors while some want a joint major that includes the social sciences and others are majoring in the humanities and social sciences and are attracted to our class out of interest as well as the practical advantage of a second Science course on their transcript. Our combined scholarly areas of expertise and personalities as well as our different life experiences showcase to students the benefit of intellectual collaboration.

---

3 "Leslie Gregg-Jolly, Professor of Biology; Grinnell College, Iowa.
4 These are the routine abbreviations for Doctor of Medicine, Master of Business Administration, and Doctor of Philosophy.

Our course is titled *Racing Through Genetics* which spotlights "race" as an action verb. The major challenge is to discredit false ideas about immutable human differences which students have inherited and internalized from the world in which they – and we – reside. Most public conversations about racism in general, and health disparities in particular, speak and think about groups of people with the vocabulary of races as categories. This discourse presumes that "race" is self-evident which is reinforced by checking boxes on forms that ask individuals to identify with "a racial category". The purpose of collecting population data is ostensibly for positive purposes. I will return to this theme in a few moments.

Our approach promotes and insists on asking and reiterating: what difference does difference make? The question helps anchors an appreciation for differentiating "good science" from "bad science" which, as my biology colleague repeatedly reminds students, is that: good science is rife with surprises based on looking at the data upside down and backwards; bad science glances at the data with an eye a to confirm preconceived.

Although our course is designed for undergraduates, we are pushing toward a long-term objective to bring the insights to colleagues at our institution and elsewhere. The main challenge we encounter is disentangling "misperceptions" about human difference and socio-politically defined demographic categories among Biology professors and health professionals. Because students are educated by professors and graduate to become researchers, professors and professionals – we explicitly recognize that science syllabi usually do <u>not</u> include material from social sciences and humanities. This serious omission must be addressed if biologists, geneticists, and other scientists are to understand the role that science has played in the construction of race classifications and how these racial classifications continue to impact research, diagnosis and treatment in ways that are misleading and even harmful. One key intervention strategy is to "historicize" and "contextualize" biomedical and genetic research "as a prerequisite" for becoming attentive to the trace of scientific racism that percolates into the medicalization of race in pharmaceutical, medical and genetic research. This approach clarifies the problem of "prescriptive thinking" when it comes to deeply internalized notions of race-based difference which bias research questions and interpretations.

Repeatedly we hear students and faculty comment "race is a social construction" and then retract by speaking about race as if it were possession (to have a race) – a biological legacy they have inherited and inhabit. The over-used phrase "social construction" has become an aphorism which actually overshadows the most significant point: social categories are named by scientists, institutionalized by politicians and policy-makers, expounded on by intellectuals and normalized in everyday conversations and practices.

Race is a classificatory system and classifications require criteria. All schemas of race categories rely on selective ancestry as their criteria and necessarily bracket inconvenient contradictions as exceptions. Race is not a salient noun – you cannot buy it at the grocery or pick it up at the local pharmacy. Race is – quite simply – a verb, a predicate, an action.

> Racism is first and foremost a social practice
> Racism always takes its principal unit and core concept "race"
> For granted as objective reality.
> BUT
> "Race" is NOT an objective reality
> Racism IS a reality

3 Historicize and contextualize

In the Americas, people are still being raced under variations of three abstract aggregate labels of white, black, mixed. In contrast, many countries in Europe exclude "Race" as a distinct category in their population data while using "Ethnic" or "National Origin". The aversion to spurious race categories reflects the pernicious echo of biological difference intrinsic to Nazi scientific racism culminating in the extermination policies against Jews demonized as an inferior race. Where researchers educated in Europe are discouraged from using "race" as "an analytic unit" in their demographic surveys and population studies, the taboo is disregarded when they join U.S. American peers and publish or deliver conference papers in English.

Europeans are, let us be clear, also facing their reckoning with racism and colonialism and the import of American-made theories has also forced a conversation about language and representation – what vocabulary to use for groups of people who have been subject to systemic racism. Parenthetically, conceptualizing difference and discrimination based on ideas of immutable difference should become a global project which includes comparative studies with, for example, discussion about Japanese racism toward Koreans, Chinese racism toward Vietnamese, and so forth.

Let me be more precise about history and context with specific details. In the late 17th century, British colonists who settled across the Atlantic realized the need to define the status of slaves and the criteria for permanent bondage because enslaved women raped by English men became pregnant and gave birth to babies. The question was: are these babies free or slave? Virginia was the first colony to address the question in the context of a legal definition for property; the child follows the status of the mother:

> Whereas some doubts have arisen whether children got by any Englishman upon a negro woman should be slave or free, be it therefore enacted and declared by this present Grand Assembly, that all children born in this country shall be held bond or free only according to the condition of the mother; and that if any Christian shall commit fornication with a negro man or woman, he or she so offending shall pay double the fines imposed by the former act (Act XII, Laws of Virginia, December 1662, Hening, Statutes at Large, 2: 170)

The other colonies very quickly followed the Virginia precedent. Color of the child was not included in legal definition and only status of the mother was inherited by the child. Moreover, the literature – fiction, autobiographical and news reports – provide testimony to the unreliability of skin color for distinguishing free and slave or as racial criteria. For example of false premise of scientific race as biological, one need only look to the abolitionist advertising campaign displaying dark brown and very white enslaved children as poster children to mobilize sympathy – and condemn adultery on southern plantations as well as the autobiography of Harriet Jacobs, whose real name was Linda Brent *Incidents in the Life of a Slave Girl* (1861) and Mark Twain's novel *The Tragedy of Pudden' head Wilson* (1893). Color was not a significant "racial" criterion until the abolition of slavery and involuntary servitude in 1865 (US Constitution 13th amendment).

> Which child here is "Black"?
>
> How do you know? Which child here is a "Person of African descent?" How do you know? Which child here is a "Free Person?" How do you know?

Let us pause over a time frame: the American colonies become independent in 1776 officially accommodating the contradiction between a rhetoric of freedom and the practice of slavery. By the end of the 18th century, trade in human beings across the transatlantic route and within the Americas was profitable for wealthy English, Spanish, British, and Americans. Consider the moment: the French were overthrown in St. Domingue by African-descent slaves in the Haitian Revolution between 1791–1804; German naturalist Johann Blumenbach was creating categories for human variation while German philosopher Georg W. F. Hegel, was philosophizing about the master and the slave (Susan Buck-Morss 2000). Obsession about human variation and classification would prove instructive when folded into a taxonomy of race which became foundational to a professionalized Science of Race married to Anthropology (the study of man), Anatomy and Medicine. These are not arbitrary snippets from the past but pivotal to comprehending

Isaac & Rosa, Slave Children from New Orleans.
Photographed by Kimball, 477 Broadway, N. Y.
Ent'd accord'g to act of Congress in the year 1863, by Geo. H.
Hanks, in the Clerk's Office of the U.S for the So. Dist. of N. Y.

Fig. 1: Isaac & Rosa, slave children from New Orleans.
M.H. Kimball, Albumen print on carte de visite, c. 1863. Credit Library of Congress.

how ideas about immutable, inherent characteristics become a fixed and ideological lens through which to conceptualize and perceive the world and conclude humans are not one that humans are not one human race but rather a hierarchy of different species.

Scientists used the authority of science to provide theoretical artillery that rationalized slavery and legitimized inequality. And therefore we take time, in our class, to note that no consensus ever emerged "on how many races existed" within the human species and corollary debates over racial purity were never really settled. The debates themselves are evidenced, sometimes with ridicule, in judicial deliberations in cases over what race category to assign or re-assign an individual[5].

This sprint through the past highlights selective glances at racist thinking in a time when the "science of race" was not just American. On the contrary, racial taxonomy is not partial: everyone is included and until 1924, Europe was composed of a continent of races and race, per se, registered an idea about blood and belonging. William Z. Ripley's *Races of Europe: A Sociological Study* was widely familiar and further normalized the obviousness of grouping people into "races" (Ripley 1899). Races were graphically illustrated, in caricature and political satire, evidenced by browsing, for instance, in the high-brow Harper's magazine that circulated widely in Northern cities in the U.S. Without looking at actual illustrations, or reading an unedited judicial opinion, a newspaper editorial or literary text, the effort to immerse oneself momentarily in the mindset of diverse people whose lives were intersecting, sometimes intimately, throughout the 19th century is very difficult. Trying to use current political vocabulary of ethnicity and ethnic groups does not just miss the point – it actually distorts the reality of an earlier era. Whatever social rank, political standing, or economic status, the world was literally seen differently than the ways we see today.

One mystery should be solved: we can map the timeline for the shift from a science of race to ideological scientific racism and the search for essential, immutable biological differences. As professors, we are determined that students explicitly situate their knowledge and are attentive to their learning as a process; to pay attention to why information and analysis may seem persuasive or irreconcilable when it contradicts their certitude.

For example, the confidence that race "is" – too firmly set in people's minds – is challenged when we assign students to look up and report back on

---

5 There are many cases but two which are interesting are relevant in today's discussions about immigration and naturalization, are *Ozawa v. United States260 US 178* (1922) and *United States v. Bhagat Singh Thind*, 261 U.S. 204 (1923).

US Census categories in every decade since the first in 1790. Are the categories consistent over time and what are the changes? And how are the categories named? When do changes take place: confronting the archive – not an interpretation or abridged version – reveals that white person always appears on the census, the census has not always used "race" in their taxonomy and "white" was a contested category defined by what it is not in a series of acrobatic exclusions: consider the Racial Integrity Act, Virginia 1924

> For the purpose of this act, the term "white person" shall apply only to the person who has no trace whatsoever of any blood other than Caucasian; but persons who have one-sixteenth or less of the blood of the American Indian and have no other non-Caucasic blood shall be deemed to be white persons. All laws heretofore passed and now in effect regarding the intermarriage of white and colored persons shall apply to marriages prohibited by this act (Virginia State Registrar of Vital Statistics responsible for keeping the records)

Notice the language here: Caucasian is incorporated in this act as a term with scientific value used for white persons in this new South determined to police white racial purity. But the same year, 1924, millions of Eastern and Southern Europeans had massively settled in the North where industrial jobs were available, and were perceived as racially inferior to Anglo-Saxon whites, Caucasians. Today, the presumptive science term from naturalist Blumenbach is liberally deployed by those who willfully or naively do not know its history. In contrast, 1924 was the year the anti-immigrant lobby finally succeeded in putting a break on the free entry of the "degenerate races of Europe". It goes without saying that the same verbal and cartoon demonization of immigrants from Europe to the United States – including reference to physical differences – frequently mirrors behavior of today's anti-immigrant campaigners.

Scientific racism also championed class prejudice and this is particularly evident in the literature on racial difference in European fields of ethnology and anthropology. For the US, we have our students read about poor whites who were seen as discardable and racially inferior. Our students read an excerpt from Francis Galton, who coined the term "eugenics" in 1883 to define "the science of improving stock i.e. the race" followed by an infamous judicial decision which upheld involuntary sterilization for those deemed inferior. The case of Carrie Buck offers a window into how bad science supports racist and discriminatory policies. Reading the 1927 Supreme Court decision in *Buck v. Bell* in which Justice Oliver Wendell Holmes upheld the sentence of sterilization with the words: "Three generations of imbeciles are enough". They are struck by the callousness of the decision and disturbed that it was imagined as a progressive decision[6].

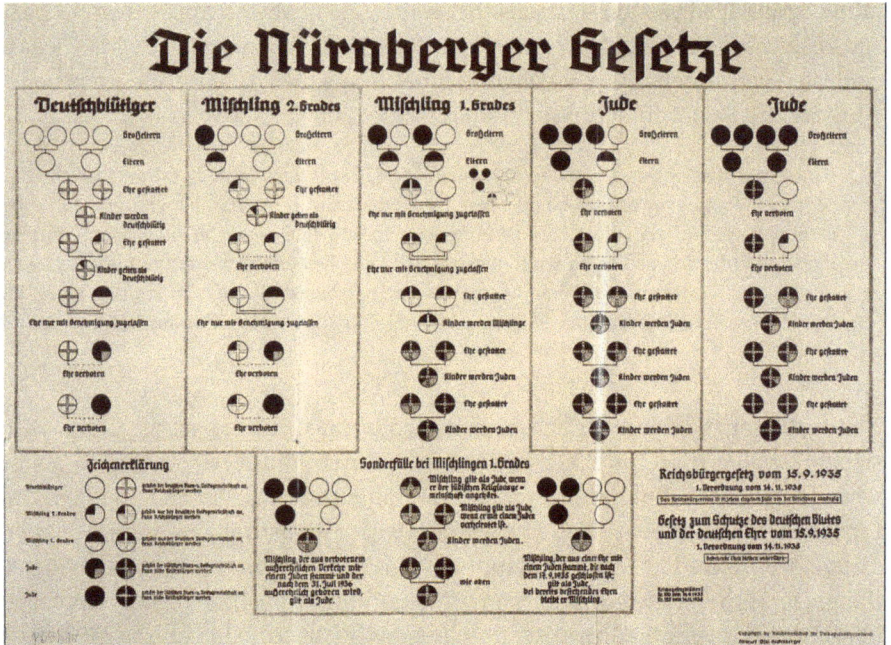

Die Nürnberger Gesetze

Fig. 2: The Nuremberg Laws.
Wikimedia Commons. https://it.m.wikipedia.org/wiki/File:Nuremberg_laws_Racial_Chart.jpg.

Eugenics of course leads directly to Nazi scientific racism and their genocidal policies towards Jews about which, many of our students know very little and cringe when we show actual images because the past should be remembered in its explicit evil. Race charts and scientific racism cannot be made more explicit – the Nuremburg Laws makes it clear that race is about ancestry and not color, about inherent difference found in the body, not in culture or socialization.

To reinforce this point, we have students look at material about science and the Holocaust and supplement their findings with Alain Resnais' 1956 film *Nuit et Brouillard*. The message in the text resonates with most students even as some of them are shocked at the images and being asked to be second-hand witnesses to the unforgiveable and the unforgettable. This half-hour film confronts students and colleagues with short memory of the irrelevance of white skin as cri-

---

6 Buck v. Bell, 274 U.S. 200, (1927)

teria for belonging to the white race category. Today Americans are having a moment of racial reckoning and the focus on "whiteness and privilege" overshadows attention to poverty, geography and the impact of systemic economic discrimination and disparity. At the same time, the crisis of COVID-19 and its
mutations has highlighted both the politics of inequality and simultaneously
the unreliability of research which uses backdoor race categories disguised as
"population genetics".

I argue here that while social science theories have pushed open meaningful
debates about racism, class and gender, these theories can also mislead and mis-
represent "cause" for "effect". When this seeps into biomedical research, what is
misleading is a health hazard.

Having established a framework for students to map the history and impact
of racing people in the United States, we turn to other countries and ask students
to look at population statistics from different countries in different continents.
The base of comparison is the United States and vocabulary is important:
there are four federally mandated race categories and one ethnic category (His-
panic/Latino)[7]. Our assignment lets students learn about very different ways in
which governments identify and classify their population. It is as much a learning exercise for international students as for those raised in the US precisely because categories are most often taken for granted by the public for whom they
are commonplace. Familiar race-informed social science categories used in the
US are poor search terms outside the US, where ethnicity, national origin, sometimes language, religion, and sometimes no population data are the organizing
units of analysis.

Does this mean that racism, xenophobia, prejudice and discrimination are
absent? Absolutely not. Should we continue to utilize social categories as a
tool to monitor discrimination? We should use social categories because they
provide one effective way to establish comparisons for differential treatment
without which it is more difficult to prove discrimination.[8] But we can no longer

---

7 In 2020, the Office of Management and Budget, responsible for the content and explanations
of the U.S. Census, instructed: OMB requires that race data be collected for a minimum of five
groups: White, Black or African American, American Indian or Alaska Native, Asian, and Native
Hawaiian or Other Pacific Islander. OMB permits the Census Bureau to also use a sixth category -
Some Other Race. Respondents may report more than one race.
8 On the efficacy of continuing to use racial categories for explicit purpose of identifying racist
discrimination and effective strategies for remedial intervention, see Katya Gibel Azoulay "Interpreting the Census: The Elasticity of Whiteness and the Depoliticization of Race." In Racial Liberalism and the Politics of Urban America eds. Curtis Stokes and Theresa Melendez (East Lansing: Michigan State University Press, 2003), 155-170.

fetishize them precisely because in the era in which we are living, intersectionality is more prominent than ever – are we differentiating by color, weight, age, presumptions about gender, religiosity, class characteristics? Physical features matter but racial designations are insufficient and opaque.

Having noted the arbitrary criteria used to group people, the question remains: if countries have no "race categories," and therefore do not collect "data on race", what criteria do their researchers use for grouping people in biomedical and genetic studies? What happens when population labels travel from one academic circle to another packaged as reliable and exported into public discourse? How do group labels translate in the laboratory by the people who carry out biological research? Precisely because science is a collaboration effort, when individuals who are socialized in different environments, educated in different schools and countries come together as research teams there is necessarily an exchange of ideas. What happens in the laboratory or the clinic is affected by what happens outside.

4 Racism, science and curating a syllabus

My argument – and the logic both I and my colleague contribute to our syllabus – is that in the area of medical research, diagnosis and treatment categorization of human beings that infer racial taxonomy, racial categories and racial derivatives should be relegated to the historiography of scientific racism and viewed as "bad science". Human variation based on subjective and learned differences are rooted in histories – political, economic, legal. These histories definitely have biological consequences which impact the health and well-being of humans. Therefore examining the legacies of discriminatory racist legislation and policies which become embedded in society and remain consequential generations after they have been abolished is a most important task. For this reason, while researchers examining health disparities need deliberately to take into account the legacy and impact of institutionalized racism, they must also commit to drafting new data schemes in research projects about people. And scientists need to unlearn the biases of the societies in which they live and refuse the biases in the places they move to work – and this is often difficult. Consider, as an analogy, the impossibility of pulling up a rug if you are standing on it[9]. The clas-

---

9 Here I am appropriating a comment by Barbara Fields, in Daniel Denvir, "An Interview With Barbara J. Fields Karen E. Fields" 17 January 2018 https://www.jacobinmag.com/2018/01/race craft-racism-barbara-karen-fields

sificatory schema of race is the rug – it is familiar, unthreatening, and so predictable. But it is no longer the best tool to fight racism, xenophobia, nationalism, or fundamentalism.

Research questions about health and illness of human beings need precision, not prescriptive investigations. Prioritizing attention to the political socio-economic variables that differentiate, discriminate, and have negative impact on the health and well-being of individuals and communities should take precedence over personal and collective identities.

Consider the calamitous impact of finding stem cell and organ donors according to identities: this is where bureaucratic classifications of "race", and its sanitized cousin "ethnicity", which use culture as a metonym for race, dangerously intersect with focused attention on genetic ancestry carried out in the commercially competitive fields of research and recreation. The treatment for the worst cases of blood cancers is stem cell matching. One would think that donor and recipient would be matched through a random computer search and it is – for some people. The problem is that individuals feed the computer based on how they index groups of people. Where race categories are used – and these are always based on selective ancestry – the outcome is predetermined to be based on race. So when donors and recipients are matched according to race categories that are re-branded as "ethnic categories", the chances for finding a match are already hampered by essentialist ideas of biological difference and misguided understandings of statistical probability and genetic research. The prevailing myth that "mixed race" people need their own registry is as egregious and evidences the longevity of racialized ideas woven into ideas about genetic distinctiveness. A misguided noble cause – what human being is *not* mixed with known and unknown descendants from around the world?

5 The challenge: Politics, policies and willful ignorance

The tenacity of racial theories of difference and willful refusal to differentiate social fact from biological fiction is infuriating precisely because a search through the science literature using the keywords "race" and "ethnicity" produces volumes of articles on research projects currently underway. One reason for this might be attributed to "vain virtue" – self-congratulation for quick solutions which should address inequality but in fact "reinforce" essentialist ideas about identity. "Identities" are always plural, political and public. Identities are not inscribed into our DNA except metaphorically and therefore regardless

ODDS OF FINDING A MATCH BASED ON ETHNIC BACKGROUND

23%	41%	46%	57%	77%
Black or African American	Asian or Pacific Islander	Hispanic or Latino	American Indian and Alaskan Native	White

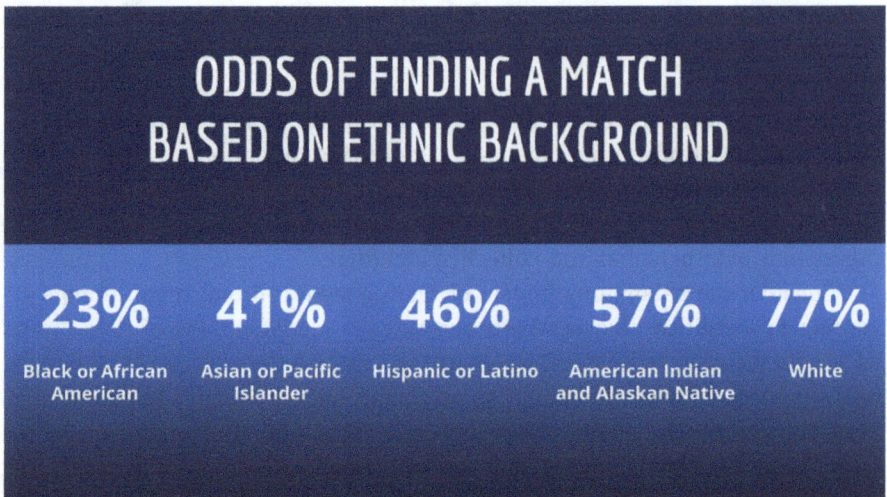

Fig. 3: Likelihood of finding a matched adult donor on the *Be The Match Registry*. Source of data: https://bethematch.org/foraj/.

of how respondents identify or are identified by bureaucratic categories, researchers should interrogate the efficacy of using racial classifications or their scientific euphemisms. But at least in the US, this is complicated by the ways scientists have interpreted The *Minority Health and Health Disparities Research and Education Act of 2000*. This bill and its adjustments over time was intended to address inequality and omissions – the directive was explicit: "to conduct and support research, training, dissemination of information, and other programs with respect to minority health conditions and other populations with health disparities".

The intent of the *Minority, Health and Disparities Act* was indeed to intervene against discriminatory omissions in research. However researchers seeking funding have misinterpreted it as an invitation to reinscribes essentialist racial categories in medical research – embedded in this mandate is the magic of racecraft which presupposes and is premised on genetic variation between pre-defined racial and ethnic minority populations: "[t]he Secretary shall ensure that any current, proposed, or future research and programmatic activities regarding genomics include focus on genetic variation within and between populations, with a focus on racial and ethnic minority populations, that may affect risk of disease or response to drug therapy and other treatments, in order to ensure that all populations are able to derive full benefit from genomic tests and treatments that may improve their health and healthcare"[10].

In other words, researchers seeking funding have to comply with the directives of the Act, and this has been interpreted as identifying groups according to the definition set out by the term "minority group" has the meaning given the term "racial and ethnic minority group" in section 1707 of the Public Health Service Act (42 U.S.C. 300 u-6) (as amended by section 501). The groups are based on the US census which is mandated by the Office of Management and Budget – given the way funding is allocated and the politics of drafting proposals which need to promise that the research will benefit communities identified as "Minorities", it is not difficult to see the incentives in reproducing racialized ideas about populations. What is needed, instead, is a reconceptualization of populations from the outset and the "only" way to do this is to make clear the difference between socio-political differences and physiological differences. Let me conclude with two examples of willful ignorance – the 1st displays the medicalization of race on labels of over-the counter medications for osteoporosis which target Asian and Caucasian women and the second is are search project that evaluates baby and toddler recognition of "racial" differences in their reaction to female adult women with black and white faces.

I hope readers cringe at this second example – but note that this was a collaborative international project carried out by a white Anglo-American Protestant from the Midwest and a Chinese researcher in China. Because the application for funding was submitted in the U.S., the proposal was drafted to meet their understanding that race – not racism – had to be central to the proposal and project. Nothing in the proposal hinted at an understanding of how all babies, everywhere and in all times, are socialized to notice familiar and strange or why the researchers presumed in advance that racial difference and skin color were the same notions. A search through science publications showed students that "criteria" for who was included in groups identified for study were rarely outlined and racial classifications were simply taken for granted as self-evident[11].

---

10 See the *Minority Health Improvement and Health Disparity Elimination Act*, S.1576–110th Congress (2007–2008), https://www.congress.gov/bill/110th-congress/senate-bill/1576/text?r=81&s= 1#toc-d117C978C30544CA D857682978023B55F, visited on 28 June 2022.
11 The title says it all: Kelly, David J.; Quinn, Paul C.; Slater, Alan M.; Lee. Kang, Gibson, Alan; Smith, Michael; 1 Ge, Liezhong; Pascalis, Olivier. (2005): "Three-month-olds, but not newborns, prefer own-race faces". In: *Developmental Science* 8. N. 6, F31–6. doi:10.1111/j.1467–7687.2005.0434a.x.

We are living in a moment when a pandemic put the world on pause and has forced medical researchers and geneticists to innovatively figure out remedial interventions. The same year that the *Minorities and Health Disparities Research and Education Act* passed, British sociologist Paul Gilroy made an appeal for liberation from "racialized seeing, racialized thinking and racialized thinking about thinking as the only ethical response to the conspicuous wrongs that raciologies continue to solicit and sanction" (Gilroy 2000). We are not yet post COVID-19 and many countries remain either in partial lockdown or lockout – Gilroy's appeal remains an urgent and relevant insight.

References

Buck-Morss, Susan. (2000): "Hegel and Haiti". In: *Critical Inquiry* 26. No 4, pp. 821–865.

Di Angelo, Robin. (2018): *White Fragility: Why It's So Hard for White People to Talk About Racism.* Boston: Beacon Press.

Fanon, Frantz. (1952): *Peau noire, masques blancs.* Paris: Éditions du Seuil.

Fanon, Franz. (1977): *Black Skin, White Masks.* New York: Grove Press.

Fields, Barbara. (2018): "An Interview With Barbara J. Fields Karen E. Fields", 17 January. https://www.jacobinmag.com/2018/01/racecraft-racism-barbara-karen-fields, visited on 28 June 2022.

Fields, Karen J.; Fields, Barbara. (2012): *Racecraft: The Soul Of Inequality In American Life.* London: Verso.

Gibel Azoulay, Katya. (2006): *"Reflections on 'race' and the biologization of difference".* In: *Patterns of Prejudice* 40. N. 4-5, pp. 353-379.

Gibel Mevorach, Katya. (2007): *Reflections on 'Race' and the Biologization of Difference. In Race and Contemporary Medicine.* Ed. Sander Gilman, London: Routledge

Gibel Azoulay, Katya. (2003): "Interpreting the Census: The Elasticity of Whiteness and the Depoliticization of Race". In: Curtis, Stokes; and Theresa, Melendez (Eds), *Racial Liberalism and the Politics of Urban America* pp. 155-170. Curtis Stokes and Theresa Melendez East Lansing: Michigan State University Press.

Gilroy, Paul. (2000): *Against Race: Imagining Political Culture beyond the Color Line.* Cambridge, MA: Belknap Press of the Harvard University Press.

Jacobs, Harriet Ann. (1861): *Incidents in the Life of a Slave Girl.* North Carolina, New York: Thayer & Eldridge.

Jacobson, M. F. (1998): *Whiteness of a Different Color: European Immigrants and the Alchemy of Race.* Cambridge, MA: Harvard University Press.

Kelly, David J.; Quinn, Paul C.; Slater, Alan M.; Lee. Kang, Gibson, Alan; Smith, Michael; 1 Ge, Liezhong; Pascalis, Olivier. (2005): "Three-month-olds, but not newborns, prefer own-race faces". In: *Developmental Science* 8. N. 6, F31–6. DOI:10.1111/j.1467–7687.2005.0434a.x.

Onishi, Norimitsu. (2021): "Will American Ideas Tear France Apart? Some of Its Leaders Think So". In: *New York Times*, October 11. https://www.nytimes.com/2021/02/09/world/europe/france-threat-american-universities.html, visited on 28 June 2022.

Ripley, William Z. (1899): *Races of Europe: A Sociological Study: A Selected Bibliography of the Anthropology and Ethnology of Europe.* New York: D. Appleton and Co.

Roberts, Dorothy E. (2011): *Fatal invention: how science, politics, and big business recreate race in the twenty-first century.* New York: New Press.

Twain, Mark. (1893): *The Tragedy of Pudd'nhead Wilson.* Hartford, Connecticut: American Publishing Company.

Cláudio Lorenzo, Marcia Mocellin Raymundo

The Whiteness of Bioethics and Racism in Healthcare

Abstract: Bioethics, as an interdisciplinary field, exhibits sufficient characteristics to allow it to address various themes associated with human relations, including issues of racism. However, the debate concerning racism in the context of bioethics is as necessary as it is urgent, particularly in view of the fact that this theme has been extremely neglected in this field. The main objective of this reflection is to encourage the field of bioethics and its practitioners to address the issue of racism transversally, not only in the theoretical field, but primarily in the context of practical and everyday ethics. With this in mind, we discuss some examples of structural and institutional racism in the field of health care, seeking to establish relationships with bioethics and human rights.

> And black people understood that great winners lift themselves up beyond the pain.
> Everything arrived, surviving on a ship. Who discovered Brazil?
> Black people faced the cruelty before them
> And still produced miracles of faith in the far west
> Caetano Veloso[1]

1 Introduction

The debate on racism in the context of bioethics is as necessary as it is urgent, particularly in view of the fact that this theme has been extremely neglected in our field. And, precisely since it is concerned with interdisciplinary connections that propose to bridge the gap between science and the humanities, it is left to bioethics practitioners to include discussion of racism transversally in their different approaches, both with respect to bioethics in general and in the context of specific themes. Or rather, it is not enough for the field of social bioethics to be concerned with this approach; it is also necessary to broaden the discussion as far as possible in the context of bioethics. In this way, the main objective of this

---

1 Extracted from: "Milagres do Povo", Caetano Veloso. Rio de Janeiro: Universal Music Group, 2008. CD (67 min), original quotation in Portuguese.

Cláudio Lorenzo, University of Brasilia
Marcia Mocellin Raymundo, Clinical Hospital Porto Alegre

https://doi.org/10.1515/9783110765120-009

discussion is to encourage bioethics practitioners to address the issue of racism transversally, not only in the theoretical field, but primarily in the context of practical and everyday ethics, thus contributing to a more just, egalitarian, and equitable society.

Racism is a factor that inflicts a great deal of pain (Carneiro 2011, p. 63). It is associated with the ancestral pains of a kidnapped and enslaved people; everyday pain throughout the course of personal life, such as the moral pain of humiliation, discrimination, and harassment; psychic pain resulting from stress, revolt, and grief; physical pain due to police aggression and firearms that result in the extermination of black children and youth or to aggression against and the rape of black women in shameful numbers, as we will see later in this chapter (Cerqueira 2021, p. 49). These are pains that white men and women do not experience socially. That is why we kindly and in solidarity ask for permission to address the pain of the other, and if we do so, this action is performed out of empathy for the struggle to promote equality and not out of any pretense to speak for or in place of those who suffer these pains. We are aware that all the advances that have been – if not yet with respect to a reduction it racist practices, at least regarding the acknowledgment of their existence – are achievements made by the struggles of black people and are not the fruit of our contributions as whites.

Thus, our way of asking for permission is to announce what our locus of enunciation is, as the Brazilian philosopher and feminist Djamila Ribeiro teaches us:

> It is fundamental for individuals who belong to the social class that is privileged in terms of social *locus* to be able to see the hierarchies produced based on this place, and how this place directly impacts the constitution of the places of subalternized groups (Ribeiro 2020, p. 85)[2].

As white Brazilians, we experience all the privileges of the whiteness that was historically constructed from the diaspora of slavery. Even though we manage to acknowledge our place in the matrix of domination and seek to free ourselves from these privileges and engage with the social struggles of black people, many of our privileges are structural and do not depend on our subjective dispositions to be included in our life stories.

We live in a country in which the majority (56%) of the population is black (IBGE PNAD 2020). The great majority of genuinely Brazilian intellectual, artistic,

---

2 Translated from the original in Portuguese.

or cultural production is directly or indirectly influenced by the African cultural legacy (Munanga 2016, pp. 92–95).

The authors of this text are aware that neither attempts to resist our privileged social *locus* (Ribeiro 2020, p. 85) nor the identitarian bonds that we may have developed with black people in our country are capable of annulling our privilege or legitimizing our speech. Only the content of the reflection that we propose can grant legitimacy to a text that proposes to fight racism and promote equality.

That being the case, we reiterate the primary objective of this reflection: to enjoin bioethics and its practitioners to reduce their negligence in relation to racism.

2 Bioethics and racism

Bioethics – considered to be an interdisciplinary field due to its intercultural attitude and its purpose as an "ethics of life" – exhibits sufficient characteristics to allow it to address various themes concerning human relations, including issues of racism. Unfortunately, this subject has been systematically neglected in the field of bioethics. However, in recent years, some important academic journals have dedicated space to the subject, as we will see.

Although bioethics has historically become recognized for its approach to issues related to biomedical advances, which have placed greater emphasis on individual interests, some issues related to collective interests are of the utmost importance in this sphere. Diversity, citizenship, and the environment are examples of issues that can – and should – be approached from a bioethical perspective. In this sense, bioethics today faces many challenges, including different aspects of racism. Bioethics and human rights are essential references with respect to the task of to addressing issues that are related to human dignity and the life sciences, as they can not only improve our understanding of the potential conflicts or dilemmas that may arise but also propose an appropriate approach to their resolution (Oliveira 2011, p. 89). This approach can improve our understanding of different attitudes regarding the same facts, since such divergences are common to all human beings, such as in the case of perceptions of birth and death, whose specificities vary across different cultures.

The fact that there are differences concerning these phenomena, which exhibit particularities in each culture, does not mean that one approach is right and the other wrong but merely that there are differences. Although seemingly simple to understand, many healthcare professionals have great difficulty understanding and respecting other cultures than their own (Raymundo et al. 2011, pp. 325–330). It is precisely in this kind of situation that bioethics, combined with the framework of human rights, can contribute to conflict resolution, mainly due to its ability to support the coexistence, citizenship, and human rights policies that are required by interculturality and to encourage respect for different world-views.

Paraphrasing Kayhan Parsi (Parsi 2016, pp. 1–2) from Loyola University, "bioethics has been unbearably white". Latin American and Brazilian bioethics – even though it has been recognizably more sensitive to the ethical conflicts that have been generated in contexts of vulnerability and social exclusion – do not escape this whiteness. While writing this chapter, when cross-referencing the "bioethics" descriptor with the "racism" or "racial prejudice" descriptor in SciELO (Scientific Electronic Library Online), the database of Brazilian academic journals, only two articles were found. In both, the issue of racism was present but not as a central theme. In the LILACS (Latin American and Caribbean Health Sciences Literature) database, the same cross-referencing found 38 articles, though in 36 of these articles racism was only mentioned without being a central theme. Finally, in the PubMed database, a similar search conducted in January 2022 without date limits found just over a hundred articles, half of which directly aimed to reflect on – or analyze – some form of racism against blacks, indigenous peoples, Gypsies, or Jews. Less than 10 % of these articles examined specifically the neglect of racism by the scientific and philosophical literature in the field of bioethics.

The problem has gained so much importance that in 2016, the *American Journal of Bioethics* dedicated a special supplement to discussing the invisibility of racism in American bioethics. The title of the associated editorial, written by the Kayhan Parsi, was "The Unbearable Whiteness of Bioethics: Exhorting Bioethicists to Address Racism" (Parsi 2016, pp. 1–2), which summarizes a little of the history of racism in the United States and its effect on the American Medical Association and denounces the strange silence of the discipline in the face of such serious and evident problems with racism in the country. In 2021, it was the turn of another American journal, the Journal of Bioethical Inquiry, to facilitate this discussion in a special supplement, which introduced the "Institutional Racism, Whiteness, and Bioethics" symposium.

These considerations seem to be sufficient to demonstrate the great distance that a discipline that proposes to undertake an ethical examination of conflicts in the field of health and the environment – and to propose norms that allow us to avoid or solve them – has taken from the issue of racism, which is the result of the very whiteness of its gaze. Therefore, to address the issue of racism in healthcare in Brazil, Afro-Brazilian sociological studies and public health studies can help us understand the size of this impact.

Let us take the notion of racism, as defined by Kabengele Munanga (Munanga 2006, pp. 46–57), to indicate a hierarchical classification of human groups based on physical elements, which distinguishes the dominators from the dominated and whose central criterion is not at all biological. Instead, it is a concept that is rooted in ideology, specifically an ideology that already highlights the purpose of domination (Munanga 2006, pp. 46–57). For this reason, it makes little difference that science has shown that there are not enough distinctions in the genetic code between whites and blacks to divide them into distinct races (Pena 2008, pp. 19–31). The nonexistence of any genetic foundations to define race does not change anything when the definition of race is purely socio-historical, and denying this point can end up favoring discriminatory practices.

Decolonial studies teach us that this notion of racial superiority (Lander 2005, pp. 8–23; Dussel 2005, pp. 24–32) was fundamental to the expansionist economic project of European nations since the period of the great navigations and those nations' violent occupation of the territories of Africa and the Americas (Mignolo 2005, pp. 33–49). The notion of race thus becomes central to the task of defining the oppositions between colonizers and colonized, in which context the white European race was and remains identified with the modern, with progress, and with civilization, while black and indigenous peoples were and are identified with the archaic, with stagnation, and with barbarism (Quijano 2005, pp. 107–130). For Quijano, race is a mental category of modernity, in which

> The idea of race, in its modern meaning, does not have a known history before the colonization of America. Perhaps it originated in reference to the phenotypic differences between conquerors and conquered. However, what matters is that soon it was constructed to refer to the supposed differential biological structures between those groups. Social relations founded on the category of race produced new historical social identities in America – Indians, blacks, and mestizos – and redefined others. Terms such as *Spanish* and *Portuguese*, and much later *European*, which until then indicated only geographic origin or country of origin, acquired from then on a racial connotation in reference to the new identities. Insofar as the social relations that were being configured were relations of domination, such identities were considered constitutive of the hierarchies, places, and corresponding social roles, and consequently of the model of colonial domination that was being imposed. In other words, race and racial identity were established as instruments of basic social classification. In America, the idea of race was a way of granting legitimacy to the relations of domination imposed by the conquest (Quijano 2000, p. 534).

In turn, Afrocentricity (Reis et al. 2018, pp. 102–119) is characterized as a field of thought and reflection concerning the centrality of the African continent in several aspects. It encompasses every possible topic, such as culture, epistemology, history, politics, economics, and African identity. In this way, the result of this historical process of exclusion and segregation is that racism has become structurally embedded within this new world order. Furthermore, it is not possible to understand the impacts of racism on healthcare if we do not understand what structural racism is and the other two dimensions that are derived from it: institutional racism and interpersonal racism.

Silvio Almeida (Almeida 2020, pp. 35–57) shows us that, being structural, racism organizes political, economic, and legal relations in so-called modern societies, and the state itself is responsible for hierarchizing cultural, ethnic, religious, and sexual multiplicities, criminalizing, domesticating, or stigmatizing those who do not promote the interests of the national identity. Or, rather, racism must be understood as a technology of power that operates by means of the discrimination and systemic oppression of subalternized ethno-racial groups. Therefore – although it is common in our indignation towards racist social practices to consider them to be abnormalities, such as behavioral anomalies exhibited by the state or members of society – racism is what constitutes normality itself today.

Racism establishes various forms of social interactions, pertaining to both conscious and unconscious mental processes. This is precisely what makes racism a structural and structuring element rather than merely a conjunctural phenomenon. Or, rather, if social life were a house, racism would not be its painting, its finishing, its decoration. Racism would be the cement, bricks, beams, and slats by means of which the house was built. That is why it is so difficult to confront it.

This structural dimension that establishes the normality of racism is exemplified by the fact that, although the data concerning the violence that results from racism are so expressive, there are no large corresponding social reactions. A study of racial iniquities in prenatal and childbirth care in Brazil found worse indicators of prenatal and childbirth care in black women than in white women (Loyal 2017). The study demonstrated that black women had a greater risk of being provided inadequate prenatal care and received less guidance regarding the beginning of labor and possible complications in pregnancy during prenatal care. Furthermore, although they suffered fewer obstetric interventions in childbirth than white women, black women received less local anesthesia when submitted to an episiotomy (Loyal 2017, pp. 7–8).

According to the Economic Commission for Latin America and the Caribbean (ECLAC), which is linked to the United Nations Population Fund – UNFPA – (CEPAL 2020, p. 213), in Brazil in 2018, the percentage of men of African descent between 15 and 49 years old who were victims of homicide was triple the percentage of white men, which indicates the clearly racialized pattern associated with this phenomenon. Approximately 90% of deaths by homicide among men of African descent in Brazil occur in this age group. The number of youths of African descent who were victims of homicide was quadruple that of youths who were not of African descent. This figure indicates the violent death of 61 Brazilian youth of African descent every day, that is, more than 2 deaths every hour.

According to data contained in the 2021 Atlas of Violence (Cerqueira 2021, p. 49), which is published by the Brazilian Institute of Applied Economic Research (IPEA) and the Brazilian Public Security Forum (FBSP):

> in 2019, black people, that is, black and brown people, represented 77% of homicide victims, with a homicide rate of 29.2 per 100,000 inhabitants. Comparatively, among non-blacks (the sum of yellow, white, and indigenous people) the rate was 11.2 per 100,000, which means that the chance of a black person being murdered is 2.6 times higher than that of a non-black person. In other words, in the last year, the rate of lethal violence against black people was 162% higher than among non-black people. Likewise, black women accounted for 66.0% of the total number of women murdered in Brazil, with a mortality rate of 4.1 per 100,000 inhabitants, compared to a rate of 2.5 for non-black women[3].

Although 56% of the Brazilian population is composed of blacks (IBGE 2019, p. 11), they represent the group that exhibits the greatest disadvantages with regard to the labor market, average income, and level of education, and they are furthermore underrepresented in political life (IBGE 2019, pp. 1–12). In addition, blacks represent only 16% of university professors (Moreno 2018). The naturalization of this data and information configures structural racism.

Institutional racism, which is very important in the field of health, since concrete processes of exclusion are based on it, is nothing more than the expression of structural racism in the organization and the provision of services in the daily life of institutions. Or, rather, institutional racism is characterized by systems of inequality based on race that operate in the daily life of institutions, whether public governmental agencies, private business corporations, health units, clinics, or public or private hospitals.

To exemplify the way in which institutional racism operates in the context of health, let us take two real cases as examples. One case is described in the article "Bioethics, Race, and Contempt" (Wilson 2021, pp. 13–22), which was included in the special supplement of the aforementioned Journal of Bioethical Inquiry on Bioethics and Racism. The other case is drawn from the actual professional experiences of one of the authors of this chapter. In the article, Yolanda Yvette Wilson recounts the story of Barbara Dawson, a middle-aged black woman who was having difficulty breathing and sought help in the emergency room of one of Florida's largest hospitals. Following an examination, the emergency room physicians treated her and, once they determined that she was stable, dis-

---

3 Translated from the original in Portuguese.

charged her. However, Ms. Dawson refused to leave. She was still experiencing difficulty breathing and sought to be examined further. The outcome of the story could not be worse, as Wilson recounts:

> The hospital staff responded by calling the police, who promptly arrested her for trespassing and disorderly conduct. Even after she collapsed outside of the arresting officer's patrol vehicle, the officer assumed she was faking and can be heard on the dashcam video telling an unresponsive Dawson, "Falling down like this, laying down, that's not going to stop you from going to jail." Within hours, Ms. Dawson was dead from a pulmonary embolism, a blood clot in her lungs.

Shortly after graduating in the field of medicine, in approximately 1988, Cláudio Lorenzo, while still a resident at the Hospital das Clínicas in Salvador, started to work in the emergency room of a state hospital in the city of Camaçari in the state of Bahia, which featured huge pockets of poverty as a result of the migration inspired by the hope of employment at the petrochemical complex constructed in the region in the 1970s. He remembers a young black man, approximately 17 years old, who worked as a construction worker and who visited the health unit every month with painful difficulties caused by sickle cell anemia. When he saw the young man arrive for the first time, he noticed a certain commotion in the health care team, an exchange of suspicious glances, a certain unwillingness to care for him; he had not particularly noticed this before, in such a marked way, in relation to other black people cared for at the site.

That is when the team informed him that "he is addicted to morphine"; every month, he comes here saying that he is in pain, and no medication resolves the pain until he receives morphine. He remembers that the team leader, a white doctor, approximately 50 years old, made a comment resembling the following: "on my shift, he already knows that he can scream all he wants, because we are not here to satiate addicts". The young doctor spent the rest of his shift distressed by those moans, which were heard throughout the unit, but he still did not know exactly what was happening. When he shared his distress with a fellow doctor or nurse, he was told to remain calm, since the young man was faking his distress.

Nonetheless, those groans in his memory led him to review the issue of pain crises in sickle cell anemia. It has been noted that strenuous work – and the patient was a construction worker – can increase the frequency and severity of pain crises in patients with sickle cell anemia. Furthermore, they often occur at an intensity that only responds to opioid analgesics derived from morphine. He also found that a myth of dependence on morphine derivatives was associated with patients suffering from frequent and acute pain crises of great intensity, although dependence is actually a rare event. He compiled this information, which was duly referenced in the medical literature, and took it to the team meeting.

Nonetheless, what is most interesting about the case is the fact that his own whiteness as a young doctor in the late 1980s caused him to think that this case featured only medical misconduct resulting from misinformation. Only many years later did he realize that the fundamental origin of that cruelty, which was disguised as moralizing and punitive conduct, was racism in its interpersonal and institutional forms, since white people in similar situations were treated differently. The mistreatment of this young patient with sickle cell anemia in an emergency situation is a paradigmatic example of how institutional racism can operate in the field of healthcare.

Examples such as these demonstrate not only the urgent need for a humanization of our society, i.e., the need to recognize in the other the rights that are inherent to every human being regardless of any characteristic of the other's diversity, but also the importance of bioethics as a necessary tool for training healthcare professionals, as a fundamental instrument to guide the performance of healthcare professionals based on recognition and respect for the person. As Carlos Madariaga rightly indicates, health is not possible in a socio-political, economic, and cultural context in which these rights are restricted, and the satisfaction of the right to health is one of the most powerful components of the construction of citizenship, since public policies pertaining to this matter can enable or hinder citizenship (Madariaga 2008, p. 4).

It is precisely on the basis of diversity that an "ethics of life" is constructed, that is, agreements pertaining to social organization that are indispensable for every society to organize its own existence, which leads us to conclude that "without diversity there is no society" (Gutiérrez-Martínez 2011, p. 27). Furthermore, the question of how to reconcile universal law with such an observation of diversity continually arises. The answer to this question could be that diversity should be viewed not as an element that justifies violations of rights but rather as a constitutive element of rights (Buglione 2008, pp. 25–33).

3 Bioethics, racism, diversity, and human rights

How can diversity be implemented as a constituent element of rights? How can bioethics contribute not only to this reflection but also to the task of ensuring that transformative actions result in the effectiveness of inclusive, fair, and equitable social proposals?

According to Juan Carlos Tealdi, human rights represent the greatest morality of our time due to the degree of universality that they have attained. Furthermore, in addition to expressing a morality that focuses on the intersection of different political and religious beliefs, human rights have legal force, which is why bioethics must base its reflection on this legally recognized universal morality (Tealdi 2008, p. 14). However, as Lynn Hunt rightly affirms (Hunt 2009, p. 19), although rights must be natural, equal, and universal, not even these three qualities are sufficient to guarantee them. Human rights only make sense when they acquire political content. Therefore, our focus is not the rights of human beings in nature; rather, it concerns the rights of human beings in society, that is, the rights of human beings in relation to their fellow beings. And it is precisely due to this "relational" characteristic that human rights are connected with ethics and bioethics, which are also "relational".

However, although they have been established and grounded, human rights have not been universally recognized and applied. As affirmed by Boaventura de Sousa Santos, the large majority of people worldwide are not the subjects of human rights but rather the objects of human rights discourses (Sousa Santos 2014, p. 15). Therefore, it is urgent that we transform these discourses into concrete actions to guarantee the effective application of human rights, primarily to excluded, marginalized, and discriminated populations. These guarantees cannot be made by throwing words to the wind but rather by every one of us taking concrete actions. Historically speaking, we are far behind in the task of taking responsibility and acting to break these structures that have been petrified on unacceptable, discriminatory bases.

For its part, the Universal Declaration of Bioethics and Human Rights, which was proclaimed and adopted by UNESCO in October 2005, dedicates several of its articles to the topic of respect for human dignity, equality, and fair and equitable treatment as well as to that of the right to non-stigmatization and non-discrimination, to solidarity and cooperation, and to respect for cultural diversity and pluralism (UNESCO 2005).

In this sense, it is important to highlight bioethical proposals that take this plurality into account, such as intercultural bioethics, which, according to María Jesús Buxó i Rey, not only offers space for reflection and investigative action but also opens up the broad territory of intermediation and transcultural translation for the purpose of connecting different knowledge systems and social realities and establishing participatory communities (Buxó i Rey 2007, pp. 29–45). Without a doubt, such communities must aim to dispense with inequalities, inequities, and injustices, since, in Madariaga's words, the relations among different cultures are unequal relations to the extent that the social relations in which they originate and are contained are unequal and since intercultural relations include hierarchies that reflect social hierarchies (Madariaga 2008, p. 5).

Carla Akotirene (2019, p. 19) teaches us that black feminism – based on the perspective of intersectionality – has demonstrated that race and gender intersect in the daily experiences of black women, ensuring that their experiences of racism are never dissociated from their experiences of chauvinism, which makes all these expressions of structural and institutional racism even more serious for women. According to Akotirene, the term intersectionality, which was initially coined by Kimberlé Crenshaw, aims to give theoretical – methodological instrumentality to the structural inseparability of racism, capitalism, and cisheteropatriarchy – producers of identitarian avenues in which black women are repeatedly affected by the intersection and overlapping of gender, race, and class, modern colonial apparatuses (Akotirene 2020, p. 19)[4].

We still encounter personally mediated racism, which is, in practical terms, the colonization of subjectivities by structural racism and, according to Camara Phyllis Jones (Jones 2000, p. 1212), can be defined as prejudice and discrimination, where prejudice refers to differential assumptions regarding the abilities, motives, and intentions of others based on their race and discrimination indicates differential actions taken toward others based on their race. For the author, personally mediated racism can be intentional as well as unintentional, and it includes acts of commission as well as acts of omission. It manifests as lack of respect (poor or no service, failure to communicate options), suspicion (shopkeepers' vigilance; everyday avoidance, including street crossing, purse clutching, and standing when there are empty seats on public transportation), devaluation (surprise at competence, stifling of aspirations), scapegoating (the Rosewood incident, the Charles Stuart case, the Susan Smith case), and dehumanization (police brutality, sterilization abuse, hate crimes) (Jones 2000, pp. 1212–1213).

As can be seen, there is no institutional racism without interpersonal racism, both because institutions are formed by people and because stigma is, as Erving Goffman noted, an eminently relational attribute, a mark that imprints on the other a deteriorated identity, a social devaluation (Goffman 1988, pp. 51–58).

There is a history of scientific documentation demonstrating that, in comparison to white people, black patients are more susceptible to experiencing adverse events in the context of primary care, receive less analgesia, are more often undiagnosed, and are more likely to receive potentially inappropriate prescriptions (Piccardi 2018, p. 9).

For its part, the pandemic we are currently experiencing has uncovered the racial inequalities in healthcare even more expressively. There is data available

---

4 Translated from the original in Portuguese.

to show us that black people are the ones who were most exposed to and contaminated by the coronavirus, the ones who most took ineffective drugs with risks of intoxication, the ones who died most frequently, and the ones who died the most without receiving care (ECLAC 2021; Araújo 2021, pp. 191–205).

4 Conclusions

We cannot falsely hope that bioethics, whether as a discipline or as a broader field of knowledge, can, by means of its academic performance, transform structural realities as complex as those that are responsible for racism and its harmful effects on the lives of people, social groups, and entire populations. These are clearly political achievements to be constructed historically by the performances of black social movements.

However, if we want to influence the direction of these achievements positively, it seems inevitable that we must face some concrete challenges, such as starting to influence pedagogical and political plans for primary and secondary education, including the ethical issues that are involved in the processes of stigmatization, prejudice, and racial discrimination; encouraging national undergraduate plans to include the issue of racism in a transversal manner and to stimulate students to understand these realities critically, thus training professionals who are capable of identifying and combating institutional racism; systematizing an ongoing academic production in bioethics that is founded on contemporary black thought and committed to fighting racism; establishing stronger links with black social movements and contributing by engaging in teaching, research, and extension activities; and seeking to provide both real and virtual spaces that feature open dialogue with society.

By confronting these challenges, academic spaces and bioethics practitioners may contribute effectively to the emergence of a harbinger of change in the direction of a bioethics that can become less ideologically white and more active in the fight against racism and the struggle for equality.

References

Akotirene, Carla. (2020): *Interseccionalidade.* São Paulo: Editora Jandaíra.
Almeida, Silvio. (2020): *Racismo estrutural.* São Paulo: Editora Jandaíra.
Araújo, Edna Maria et al. (2020): "Covid-19 – Morbimortalidade pela COVID-19 segundo raça/cor/etnia: a experiência do Brasil e dos Estados Unidos". In: *Saúde e Debate* 44. N. 4, pp. 191–205.

Buglione, Samanta. (2008): "Direitos Humanos e Direitos Sexuais: universalidade e especificidade em debate". In: Veriano Terto Jr, Ceres Gomes Victora, Daniela Riva Knauth (Eds.): Cadernos do NUPACS. Porto Alegre: UFRGS, pp. 25–33.

Buxó i Rey, María Jesús. (2007): "Antropología cultural y Bioética." In: María Casado (Comp.): Nuevos materiales de Bioética Y Derecho. Ciudad de México: Fontamara, pp. 29–45.

Carneiro, Sueli. (2011): *Racismo, Sexismo e Desigualdade no Brasil.* São Paulo: Selo Negro Edições.

CEPAL – Comisión Económica para América Latina y el Caribe (CEPAL), Fondo de Población de las Naciones Unidas – UNFPA. (2020): "Afrodescendientes y la matriz de la desigualdad social en América Latina: retos para la inclusión". https://www.cepal.org/es/pub licaciones/46191-afrodescendientes-la-matriz-la-desigualdad-social-america-latina-retos-la, visited on 17 January 2022.

CEPAL – Comisión Económica para América Latina y el Caribe. (2021): "Las personas afrodescendientes y el COVID-19: develando desigualdades estructurales en América Latina". https://www.cepal.org/es/publicaciones/46620-personas-afrodescendientes-covid-19-develando-desigualdades-estructurales, visited on 19 January 2022.

Cerqueira, Daniel et al. (2021): *Atlas da Violência 2021.* São Paulo: FBSP. https://forum seguranca.org.br/wp-content/uploads/2021/12/atlas-violencia-2021-v7.pdf. visited on 17 January 2022.

Dussel, Enrique. (2005): "Europa, modernidade e eurocentrismo." In: Edgardo Lander (Ed.): *A colonialidade do saber: eurocentrismo e ciências sociais Perspectivas latino-americanas.* Ciudad Autónoma de Buenos Aires: Colección Sur Sur, CLACSO, pp. 24–32.

Goffman, Erving. (1988): *Estigma: notas sobre a manipulação da identidade deteriorada.* Rio de Janeiro: LTC.

Gutiérrez-Martínez, Daniel. (2011): "Panorámica de los fenómenos ambientales y el desarrollo en la perspectiva de la diversidad". In: Raúl Calixto-Flores, Mayra García-Ruíz, Daniel Gutiérrez-Martínez (Eds.): *Educación e investigación ambientales y sustentabilidad: entornos cercanos para desarrollos por venir.* Ciudad de México: Universidad Pedagógica Nacional, El Colegio Mexiquense, pp. 27–34.

Hunt, Lynn. (2009): *La invención de los derechos humanos.* Barcelona: Tusquets Editores.

IBGE. Istituto Brasileiro de Geografia e Estatística. (2019): "Desigualdades sociais por cor ou raça no Brasil", https://biblioteca.ibge.gov.br/index.php/biblioteca-catalogo?view=de talhes&id=2101681, visited on 18 January 2022.

IBGE. Istituto Brasileiro de Geografia e Estatística. (2020): "Pesquisa Nacional por Amostra de Domicílio Contínua – PNAD Contínua". https://biblioteca.ibge.gov.br/visualizacao/liv ros/liv101707_informativo.pdf, visited on 22 January 2022.

Jones, Camara Phyllis. (2000): "Levels of Racism: A Theoretic Framework and a Gardener's Tale". In: *American Journal of Public Health* 90. Pp. 1212–1215.

Lander, Edgardo. (2005): "Ciências sociais: saberes coloniais e eurocêntricos". In: Edgardo Lander (Ed.): *A colonialidade do saber: eurocentrismo e ciências sociais. Perspectivas latino-americanas.* Ciudad Autónoma de Buenos Aires: Colección Sur Sur, CLACSO, pp. 8–23.

Leal, Maria do Carmo et al. (2017): "A cor da dor: iniquidades raciais na atenção pré-natal e ao parto no Brasil". In: *Cadernos de Saúde Pública* 33. Sup 1: e00078816, pp. 1–17.

Madariaga, Carlos. (2008): "Interculturalidad, salud y derechos humanos: hacia un cambio epistemológico". http://www.cintras.org/textos/reflexion/r36/intercularidad.pdf, visited on 19 January 2022.

Mignolo, Walter. (2005): "A colonialidade de cabo a rabo: o hemisfério ocidental no horizonte conceitual da modernidade". In: Edgardo Lander (Ed.): *A colonialidade do saber: eurocentrismo e ciências sociais Perspectivas latino-americanas.* Ciudad Autónoma de Buenos Aires: Colección Sur Sur, CLACSO, pp. 33–49.

Moreno, Ana Carolina. (2018): "Negros representam apenas 16% dos professores universitários". https://g1.globo.com/educacao/guia-de-carreiras/noticia/2018/11/20/negros-representam-apenas-16-dos-professores-universitarios.ghtml, visited on 18 January 2022.

Munanga, Kabengele. (2006): "Algumas considerações sobre "raça", ação afirmativa e identidade negra no Brasil: fundamentos antropológicos". In: *Revista USP* 68. Pp. 46–57.

Munanga, Kabengele. (2016): *Origens africanas do Brasil contemporâneo.* São Paulo: Global Editora.

Oliveira, Aline Albuquerque de. (2011): *Bioética e Direitos Humanos.* São Paulo: Edições Loyola.

Parsi, Kayhan. (2016): "The Unbearable Whiteness of Bioethics: Exhorting Bioethicists to Address Racism". In: *American Journal of Bioethics* 16. N. 4, pp. 1–2.

Pena, Sérgio Danilo Junho. (2008): *Humanidade sem raça?.* São Paulo: Publifolha.

Piccardi, Carlota et al. (2018): "Social disparities in patient safety in primary care: a systematic review". In: *International Journal for Equity in Health* 17. N. 114, pp. 7–9.

Quijano, Aníbal. (2000): "Coloniality of Power, Eurocentrism, and Latin America". Translated by Michael Ennis. In: *Nepantla: Views from South* 1. N. 3, pp. 533–580.

Quijano, Aníbal. (2005): "Colonialidade do poder, eurocentrismo e América Latina". In: Edgardo Lander (Ed.): *A colonialidade do saber: eurocentrismo e ciências sociais Perspectivas latino-americanas.* Ciudad Autónoma de Buenos Aires: Colección Sur, CLACSO, pp. 107–130.

Raymundo, Marcia Mocellin et al. (2011): "Bioética y salud intercultural: apuntamientos para una conexión necesaria y posible". In: *Rev Med Inst Mex Seguro Soc* 49. N. 3, pp. 325–330.

Reis, Mauricio de Novais et al. (2018): "Afrocentricidade: Identidade e centralidade africana". In: *Odeere: Revista do Programa de Pós-Graduação em Relações Étnicas e Contemporaneidade* 3. N. 6, pp. 102–119.

Ribeiro, Djamila. (2020): *Lugar de fala.* São Paulo: Editora Jandaíra.

Sousa Santos, Boaventura. (2014): *Se Deus fosse um ativista dos direitos humanos.* São Paulo: Cortez.

Tealdi Juan Carlos. (2008): "Bioética de los Derechos Humanos. Investigaciones Biomédicas y dignidad humana". In: *Boletín Trimestral de la Red Iberoamericana de las Libertades Laicas* 4. N. 15. http://www.libertadeslaicas.mx/images/libela/libela15.pdf, visited on 19 January 2022.

UNESCO. (2005): Universal Declaration on Bioethics and Human Rights. https://en.unesco.org/themes/ethics-science-and-technology/bioethics-and-human-rights, visited on 19 January 2022.

Wilson, Yolanda Yvette. (2021): "Bioethics, Race, and Contempt". In: *Journal of Bioethics Inquiry* 18. N. 1, pp. 13–22.

Josimário João da Silva, Ronaldo Piber
Misthanasia: Bioethical Reflections on the Cruel Face of Human Indifference

Abstract: Misthanasia is a term that suggests a miserable, early, and preventable death. In this chapter, we use this term to evoke the increasing phenomenon of miserable, early, and preventable death in Brazil, which is induced by the maintenance of some population groups in a condition of poverty and thus by a lack of the minimum conditions necessary to live a decent life. From this perspective, misthanasia does not merely coincide with physical death; rather it begins the moment in which people are denied their fundamental rights and, above all, access to the minimum conditions of citizenship. After sketching out the potential contribution of bioethics, particularly public health ethics, we suggest that reinforcing education in values can facilitate the identification of the other in a condition of vulnerability who must be accepted and welcomed in light of its fragility and above all its dignity.

1 Introduction

In contemporary society, everything progresses very quickly. There is no time to waste. People rush to maintain their "social status". It is a society based on consumer goods. There is a form of "eudaimonia" in being able to access consumer goods alongside an impression that people are happy. But the reality is crueler for those who, for social, political and economic reasons, have not had the same opportunity to experience this eudaimonia of the few. Today large portions of marginalized people live in subhuman conditions due to the structural racism that is increasingly rooted in our society.

In this chapter, we use the term "misthanasia" – which is derived from the Greek *mis* (unhappy) and *thanatos* and means "unfortunate death" – to evoke the increasing phenomenon of miserable, early, and preventable death in our country, which is induced by the maintenance of some population groups in a condition of poverty and thus by a lack of the minimum conditions necessary to live a decent life. From this perspective, misthanasia does not merely coincide with physical death; rather, it begins the moment in which people are denied

Josimário João da Silva, Federal University of Pernambuco
Ronaldo Piber, University of Santo Amaro

https://doi.org/10.1515/9783110765120-010

their fundamental rights and, above all, access to the minimum conditions of citizenship.

Due to the increasing number of people living below the poverty line, the state has revealed its inability to promote social inclusion and citizenship. In other words, the state has shown itself to be unable to provide its population with the means necessary to satisfy its basic needs. With respect to healthcare, this entails that an increasing number of patients wait for surgeries and consultations, lack adequate access to drugs and treatments, and do not receive timely diagnosis and rehabilitation.

As an example, we can mention the outbreak of the Zika virus, which left hundreds of Brazilian children with severe sequelae affecting their cognitive and motor development (Lin et al. 2018; Ornelas Pereira et al. 2020). Today, these patients depend on rehabilitation activities and, in most cases, do not receive treatment that can meet their basic needs. Due to the COVID-19 pandemic, the number of people who lack adequate access to care in Brazil has increased (Pereira et al. 2021; Cimerman et al. 2020). As a result, the number of patients affected by other diseases with declining clinical conditions has also notably increased.

2 The role of the state in defending people's health

According to the WHO (2013), health refers to complete physical, mental, social and even spiritual well-being, with full awareness of the fact that the human being is composed of both body and soul within a biological structure. For Plato, man is body and soul, the soul being what we really are and the body being merely something material that follows us. Taking this concept into account, it is clear that we, as healthcare professionals, should care about our patients rather than solely about the diseases they have. The patient must therefore be regarded as an entity who deserves to be treated according to their physical and spiritual needs. Indeed, the need to ensure quality in healthcare has become a major challenge for contemporary public health policies.

The Constitution of the Federative Republic of Brazil of 1988 (hereafter referred to as the Constitution) stipulates that health is the right of everybody and a duty of the state as well as that health should be ensured through: *i*) social and economic policies aimed at reducing the risk of disease and other injuries and *ii*) universal and equal access to the actions and services that are necessary for health promotion, protection and recovery (Constitution art. 196). In the Constitution, the right to health is listed among the social rights and included in the section "Fundamental Rights and Guarantees of Man" (Constitution art. 6). Indeed, each social entitlement is also considered to be a fundamental right, which entails that its application should be immediate. Accordingly, all Brazilian citizens – as well as all foreign citizens on Brazilian soil – have the right to access any kind of medical treatment, including diagnostics, hospitalization, surgery or the provision of vital medications for the protection and maintenance of their health.

From the perspective of the state, the historical construction of right to health as a "fundamental" right conferred on the state the task of implementing policies and programs aimed at making its provisions concrete. This entails adopting policies aimed at fostering the prevention, protection and recovery of the population's health. The Constitution mandated universality as the basic principle of the National Health Service, which is evocatively known as *Sistema Único de Saúde* (SUS). It is precisely this universality of care – including preventive, curative and emergency treatment – that marks the innovative nature of our Constitution with respect to health protection. The choice of universality was meant to eliminate the economic, social and cultural forms of segregation that had previously prevented full access to healthcare for a large portion of the Brazilian population. As noted by Dinorá Adelaide Musseti Grotti (2004), the universality principle must be viewed as a concrete token of equality.

In other words, universal access means that all the people who require access to healthcare services should receive it, regardless of their purchasing power or any formal conditions that could be imposed for their access. As Musseti Grotti explains, the Constitution clearly aims to prevent the "elitization" of the NHS and to strengthen citizenship and the inclusion the entire population by ensuring several forms of participation in the deliberative processes related to healthcare management (Musseti Grotti 2004). To defend universal access, therefore, the SUS should take into account the peculiar conditions of various user groups, which can be fundamental to the task of defending population health and preserving it from the influence of market forces. In this regard, it must be noted that the Constitution (art. 198) also regulates the operation of private healthcare services, which are intended to be integrated with the services provided by the SUS. However, any health actions and services, whether public or private, have been classified as pertaining to activities of "public relevance".

The dignity of each person is an essential principle that permeates the entire Constitution, appearing as a vector of interpretation for all its provisions. Indeed, the "right to life", as expressed by the Constitution (art. 5) has as its axiological foundation not merely the maintenance of one's own existence but also the conditions that the state must provide to the entire population. Thus, health protection plays an inexorable role at the intersection of the right to exist and the right to dignity, which are related but should not be confused.

In addition, the right to health is also included in international legal statements concerning human rights. Health protection and healthcare access are indeed recognized as essential elements for living with dignity. In this context, it is up to the public authorities to adopt policies through which to implement mechanisms that can guarantee the protection of the health of the population. On theoretical grounds, indeed, professionals should be treated with respect and dignity, and healthcare services should be able to provide all the necessary means of caring for people's health. The implementation of public health should be considered to be a priority matter for any government, given that "life" is our greatest asset.

Notwithstanding that fact, the enjoyment of the right to health is a critical issue for the Brazilian population. Government cuts to healthcare expenditures have made even more difficult the functioning of the SUS and thus access to their services by people in need of care. Given insufficient funding sources, the managers of the SUS have been forced to choose some "priorities" to pursue. As a result, access to the services provided by the SUS is problematic for large portions of the population, particularly the more vulnerable groups that encounter multiple barriers to adequate care. In hospitals, a large number of patients huddle in the corridors of hospitals, waiting for someone to take care of them, and the number of preventable deaths is increasing throughout the whole country. In many cases, healthcare professionals are working beyond their capacity limits, i.e., without the appropriate guarantees, and they are thus exposing themselves to various risks. Finally, Brazilian citizens, who are increasingly aware of their rights, seek judicial protection to ensure that their care needs are satisfied, thus shifting healthcare from the context of hospitals to that of tribunals. That is the picture of the crude reality currently experienced by our country.

3 Racism and healthcare in Brazil

On theoretical grounds, the SUS can be depicted as a complex and ambitious NHS because it aims to serve the entire population – regardless of socio-economic, cultural, or spiritual factors or beliefs – by providing universal and equitable access to healthcare and the integral satisfaction of care needs. However, if we examine the situation of our system, we cannot disregard the fact that the enjoyment of the services offered by the SUS are influenced by a range of factors such as people's color, their sexual orientation or their socio-economic conditions. All this has been already analyzed and framed under the label of "institutional racism", i. e. the "collective failure of an organization to provide an appropriate and professional service to people because of its color, culture or ethnic origin. Institutional racism can be clearly identified in processes, attitudes and behaviors that amount to discrimination by involuntary prejudice, ignorance, neglect and racist stereotypes, which causes disadvantages to people from an ethnic minority"[1] (Kalckmann et al. 2007, p. 146).

Racism occurs in contexts ranging from the prenatal follow-up of young, black, low-income and low-education women to patients who are hospitalized by emergencies, who have their care delayed because there is no political will to improve the NHS; consequently, people do not receive the treatment to which they are entitled (Oliveira and Kubiak 2019). As has been widely acknowledged (Silvério 2002), black people are more frequently exposed to a number of diseases due to the historical processes of social, economic, political and cultural exclusion to which they have been submitted (Gomes 2019). Given the inspiring principles mentioned above, the SUS should be able to offer differentiated and specific services with the aim of reducing the gaps among population groups and fighting vulnerability. Paradoxically, the way in which our NHS performs today increases the vulnerability of these groups, thereby expanding barriers to access and leading to the exclusion of these users from services. This has also been confirmed by a national survey conducted by the Brazilian Institute of Geography and Statistics (IBGE 2019), which observed that blacks have less access to medical care, receive fewer consultations with dentists, find it more difficult to obtain prescribed medications, and have less private health coverage than the white population. These indicators confirm that inequality and racism permeate the functioning of the SUS today. In other words, institutional racism therefore operates as a "filter mechanism", preventing some population groups, particularly black people, from accessing timely and adequate healthcare services, thereby causing enormous damage to their health conditions or making their healing process impossible.

4 From vulnerability to social justice: the contribution of public health ethics

The term "vulnerability" is derived from the Latin language *vulnerabilis*, which can be translated as "something that causes injury" (Houaiss and Villar 2009).

---

1 Translated from the original Portuguese.

From a philosophical perspective, vulnerability represents a condition that is inherent to the human existence in terms of both its finitude and its fragility, in such a way that it cannot be overcome or eliminated. By recognizing themselves as vulnerable, people can also understand the vulnerability of the other as well as the needs for care, responsibility and solidarity (UN 2005).

The condition of vulnerability can be linked to multiple societal causes. Among these causes are poverty, social injustice, discrimination, diseases, violence (Moratalla 2004). However, all of them can be traced back to a feeling of "human indifference" towards the other. In parallel to this is the inertia from the state, which should be committed to providing the population with public policies and concrete measures for ensuring social and economic inclusion, especially for those who are most vulnerable.

It is worth remembering that vulnerability potentially affects everyone. The human being is not only the "capable" being, the one who can; he is also the "incapable" being, the one who cannot. We all are somehow that mixture of capacity and incapacity; we all are made before we can start doing things on our own, and that is what makes us, in a sense, passive rather than active (Kottow 2003). Precisely because it is an inherent condition of the human being, vulnerability can undermine the dignity of the human person and his autonomy.

In accordance with its principles, bioethics mandates the protection of the dignity of the human person, operating at the intersection between human vulnerability and social justice. In particular, public health ethics (PHE) is the branch of bioethics that aims at identifying, analyzing and addressing the ethical problems that are inherent to population health. PHE thus embraces the analysis of population needs and the evaluation of public policies, and it is intrinsically oriented towards results. PHE aims at promoting and preserving individual and collective well-being and therefore at neutralizing factors that threaten the maintenance of such well-being. This includes acting on the social and economic determinants of health and thus also on areas that do not belong to the field of health. In light of this, it is possible to argue that PHE is concerned with social justice.

Autonomy is also a core element of PHE, given the relevance of promoting the awareness of individuals to increase their levels of agency regarding matters of health maintenance and healthcare access. From the perspective of PHE, however, the individual cannot help being part of the collective, thus highlighting the value of relations and the relational perspective. In other words, principles such as non-maleficence, beneficence and autonomy pertain to dependency relationships that are established among moral agents, both on individual and collective grounds. The final outcome of this complex set of relations is that certain individuals are positioned in situations of vulnerability. It is already well-known that most vulnerable people, particularly the poorest, are more likely to have negative experiences with medical treatments or to be exploited as research participants due to their inability to refuse enrollment (Mackenzie et al. 2014).

In addition to individual interests, PHE is concerned with assets that are crucial to the wellbeing of the whole community and that are deemed to be essential to achieving common ends as well as preventing harm among people. PHE relies on collective action, which is often carried out by the state, and it pertains to a conception of social justice that is both sensitive to and based on the reality of relevant social relations (Dawson 2011; Dawson and Verweij 2007). In such a context, analysis of social vulnerability embraces a range of factors that are able to influence both individual and the collective wellbeing. These include not only health or medical issues but also economic conditions, political situations, gender issues, discrimination and racism, social class, etc.

In conclusion, we believe that bioethics can play a fundamental role in the task of fighting racism, discrimination and cultural segregation in healthcare. In particular, PHE can serve as a "bridge" to reduce vulnerability and enhance social justice. However, to achieve this goal, a cultural shift is necessary – alongside political action – to remove the barrier of "human indifference", which entails renewing the meaning of humanity and human values. Education in values can facilitate the identification of the other in a condition of vulnerability who must be accepted and welcomed in light of its fragility and above all its dignity. From this perspective, the Brazilian society must make all possible efforts to pursue the common good and fight to enhance the equity and accessibility of our NHS, which is essential to the tasks of guaranteeing people dignified lives and preventing misthanasia.

References

Cimerman S., Chebabo A., Cunha C.A.D., Rodríguez-Morales A.J. (2020): "Deep impact of COVID-19 in the healthcare of Latin America: the case of Brazil". In: *Braz J Infect Dis* 24. N. 2, pp. 93–95. doi: 10.1016/j.bjid.2020.04.005.

Dawson, Angus. (2011): "Resetting the parameters: public health as the foundation for public health ethics". In: Dawson, Angus (ed.), *Public Health Ethics : Key Concepts and Issues in Policy and Practice*, pp. 1–19. Cambridge: Cambridge University Press.

Dawson, Angus; Verweij, Marcel. (2007): "The meaning of 'Public' in 'Public Health'". In: Dawson, Angus; Verweij, Marcel (eds.), *Ethics, Prevention and Public Health*, pp. 13–29. New York: Clarendon Press.

Gomes, Laurentino. (2019): *Escravidão – Volume 1: Do primeiro leilão de cativo sem Portugal até a morte de Zumbi dos Palmares*. Brasilia: Globo Livros.

Houaiss, Antonio; Villar, Mauro de Salles. (2009): *Dicionário Houaiss da Língua Portuguesa*. Rio de Janeiro: Objetiva.

IBGE [Instituto Brasileiro de Geografia e Estatística]. (2019): *Pesquisa Nacional de Saúde – PNS 2019*. Rio de Janeiro: IBGE.

Kalckmann, Suzana; et al. (2007): "Racismo institucional: um desafio para a eqüidade no SUS?" In: *Saúde e Sociedade* 16. N. 2, pp. 146–155.

Kottow, Michael H. (2003): "The Vulnerable and the Susceptible". In: *Bioethics* 17. N. 5–6, pp. 460–471.

Lin, Hsiao-Han; et al. (2018): "Zika virus structural biology and progress in vaccine development". In: *Biotechnol Adv* 36. N. 1, pp. 47–53.

Mackenzie, Catriona; Rogers, Wendy; Dodds, Susan. (2014): "Introduction: What Is Vulnerability, and Why Does It Matter for Moral Theory?". In: Mackenzie, Catriona; Rogers, Wendy; Dodds, Susan (eds.), *Vulnerability: new essays in ethics and feminist philosophy*, pp. 1–29. New York: Oxford University Press.

Moratalla, Tomás Domingo. (2004): "Lectura bioetica del ser humano autonomia y vulnerabilidad". In: *Thémata* 1. N. 33, pp. 423–428.

Musseti Grotti, Dinorá Adelaide. (2004): "As Agências Reguladoras". In: *R Bras de Dir Público* 2. N. 4, pp. 187–219.

Oliveira, Beatriz Muccini Costa; Kubiak, Fabiana. (2019): "Racismo institucional e a saúde da mulher negra: uma análise da produção científica brasileira". In: *Saúde em Debate* 43. N. 122, pp. 939–948.

Ornelas Pereira, Isabela; Santelli, Ana; Leite, Priscila L., et al. (2021): "Parental Stress in Primary Caregivers of Children with Evidence of Congenital Zika Virus Infection in Northeastern Brazil". In: *Matern Child Health J* 25. N. 3, pp. 360–367.

Pereira, Rafael H. M.; Braga, Carlos K. V.; Servo, Luciana; et al. (2021): "Geographic access to COVID-19 healthcare in Brazil using a balanced float catchment area approach". In: *Soc Sci Med* 273, 113773. doi: 10.1016/j.socscimed.2021.113773.

Silvério, Valter Roberto. (2002): "Ação afirmativa e o combate ao racismo institucional no Brasil". In: *Cadernos de Pesquisa* 1. N. 117, pp. 219–246

UNESCO. (2005): *Universal Declaration on Bioethics and Human Rights*. https://en.unesco.org/themes/ethics-science-and-technology/bioethics-and-human-rights, visited on 22 July 2022.

WHO. (2013): *Spiritual health, the fourth dimension: a public health perspective*. https://apps.who.int/iris/bitstream/handle/10665/329763/seajphv2n1_p3.pdf?sequence=1&isAllowed=y, visited on 22 July 2022.

Inês Faria, Laura Brito, Karla Costa

Prejudiced Rationales for Stereotyping: On the Experiences of Black and African-Descended Women in Reproductive Care in Portugal and Mozambique

Abstract: Resorting to cases and data drawn from studies conducted in two countries with a colonial relationship, i.e., Mozambique and Portugal, we want to deconstruct histories of discrimination and prejudice in reproductive care. By exploring regimes of invisibility and forms of structural violence related to black women, we argue that stereotypization rationales based on ideas of race, gender, and class that are prevalent in reproductive care settings must be overcome, and the remains of colonial history and practices of subalternization must be rendered visible. We suggest that people – as citizens, researchers, practitioners, and patients – must come together to debate, discuss and promote horizontal conversations, breaking free from taboos and engaging in cross-sectorial and multidisciplinary dialogues, even if experimental, to provoke such a change.

1 Introduction

In this chapter[1], we want to reflect on the ways in which prejudice appears in everyday interactions between health care professionals and users of reproductive care in multiple forms. Our reflection is based on the identification and discussion of episodes of discrimination and prejudice involving racialized women that occurred in the contexts of research concerning reproductive health and the provision of reproductive health care in Portugal and Mozambique.

It has been reported several times that combatting racial and/or ethnic prejudice and its consequences starts by recognizing colonial systems and their role

---

1 This research work was funded by the Portuguese Foundation for Science and Technology (FCT-MCTES) through the PhD Grant SFRH/BD/144322/2019 and the UIDP2020 at the Research Centre in Economic and Organizational Sociology.

Inês Faria, University of Lisbon
Laura Brito, University of Coimbra
Karla Costa, Iam-Fiocruz

https://doi.org/10.1515/9783110765120-011

in the rationales of prejudice. This task also requires awareness of the fact that, instead of ending with the official liberation of colonies, colonialism has endured in the form of scattered practices and modes of institutional functioning that remain globally spread today (Hamed et al. 2020). Stereotypization and prejudiced practices found fertile terrains for expansion in neoliberal and heteropatriarchal social configurations. They have, nevertheless, been increasingly problematized (Santos 2018; 2019; Kilomba 2008; Mbembe 2013). Even if Frantz Fanon long ago explained how constructions such as race, gender and class operate in an interconnected way and may mutually reinforce each other as part of a set of violent colonial mechanisms of oppression and subalternization (Fanon 1952), we must examine these phenomena today and understand where, why and how they still occur.

By focusing on everyday practices of stereotypization and subalternization related to reproduction and reproductive health that are permeated by ideas of race, gender, and class, we can attempt to recognize such remains of the colonial past in these moments (Trouillot 2003; Mbembe 2001). Examining the mundane sites in which these dynamics are rendered visible and focusing on the context of reproduction and reproductive care in Portugal and Mozambique, our proposal is thus to reflect on empirical examples from different sites – including stereotyping opinions, accounts of interactions between black women and assisted reproductive technologies (ARTs), and clinical practices – and identify episodes of racism that are related to the legacy of colonial stereotypes that continue to operate unnoticed.

2 Racism, science and biomedicine: a short theoretical background

To situate our studies and the discussion we wish to start, we feel that it is necessary to provide some background on certain constructions involving race, biomedicine and reproduction that seem to affect the realities of our research processes and interlocutors.

Camisha Russell's (2015) discussions are relevant as a starting point, particularly the idea that considering race as a construction that does things in society is much more productive for understanding racism in the context of everyday life than focusing on the (already overdone) discussion of its existence and scientific validity *per se*. Race, in connection with other elements such as gender and class has, for a long time, been included in narratives that legitimize the mistreatment or negligence of a significant part of the global population. It is important to consider the pieces of which these practices are composed and the localized socio-technical arrangements that constitute the regimes of invisibility and structural violence of which they are part (Farmer 2004; Latour 2010; Bagnol et al. 2010; Hamed et al. 2020). These regimes and the associated violence are composed of "discrete" and unremarkable sets of attitudes, beliefs, values, materialities, invisibilities, and relations consciously and unconsciously working together (cf. Girard 2019).

Regimes of invisibility and structural violence are present in everyday life in general and in the arena of biomedicine in particular (Farmer 2004; Latour 2010; Valdez and Deomampo 2019; Rapp 2019). In fact, if we examine clinical sites and the dynamics of health care provision, it is possible to identify racist and exclusionary practices mirroring the past that persist within complex organizational power systems and interactions (Hamed et al. 2020; Lock and Nguyen 2005).

Racism in the context of science has developed within the wider context of a quest to hierarchize human beings, and it was politically and ideologically aligned with racist policies developed beginning in the 18th century. During this period, colonial regimes and occupations in sub-Saharan Africa often became entangled with narratives emphasizing the beliefs that black people were passive, resistant to pain, and, consequently, resistant to physical, moral, and psychological violence – this legitimized medical negligence (Hoberman 2012). These rhetorics expanded globally, including to the multitude of places from which black people were taken to be enslaved. At the end of the 19th century, such conceptions of race had already become consistent and integrated into the Western political ideology. In this configuration, the supposed differences between white and non-white people provided justifications for violent "civilizational missions" (Havik 2013; Mbembe 2001, p. 4) in colonial territories in Africa and elsewhere.

Simultaneously, biomedicine was developed as a trial-and-error scientific discipline supported by ethically objectionable practices that were politically allowed, constructed, and mobilized to justify processes of colonial domination, racist phenotypic classifications, and segregation politics (Farmer 2004; Lock and Nguyen 2005). The latter was translated into the establishment of a symbolic and practical equivalence between the local, in this case black, populations of the colonies and degeneration (Coghe 2020). This equivalence was supported by the belief that the immunity of the local populations to tropical infectious diseases – which were associated with putrefaction and a menace to the nervous system – meant that they were physically and morally degenerate. The consequence of this and other equivalences, particularly in urban areas, were processes of spatial segregation between European citizens and locals, which were legitimized by mutually reinforcing racial and health-related constructions (Coghe 2020, pp. 5-6; Carde 2011).

Although this content would be sufficient to produce a set of many volumes, we feel that it is important to note that practices of segregation and the regimes of colonial domination were different from each other and were experienced differently in rural and urban areas. Additionally, it is also important to mention that, despite serving purposes of legitimization of colonial occupation and domination, the administrative policies and "civilizational mission" rhetorics of the 19th century mentioned above had been in development for centuries as part of complex processes that included conflicts and negotiations made in Europe and between colonial forces and local elites in different areas (Coghe 2020; with respect to the case of Mozambique, see Newitt 1995). These dynamics and the national politics of colonizing countries play an important role in the memories of colonization and racism, in the traces of colonial history that remain in contemporary interactions, and in the politics that have been developed to address them.

3 Situating the research

3.1 Research design and methods

The findings presented in this chapter stem from data collected by Inês Faria (hereafter referred to as I.F.), as part of an ethnographic research project concerning the therapeutic trajectories of women and men living in involuntarily childless relationships in Maputo, Mozambique and SaMaNe, which stands for *Saúde das Mães Negras* (Health of Black Mothers).

This research, which was based in Mozambique, included ethnographic observations in one public hospital and one private clinic in Maputo and visits to private clinics in South Africa that cared for patients from Mozambique (Faria 2015; 2016; 2018; 2021). The study was also based on the personal accounts and experiences of 25 women. This group included: five women who had undergone reproductive travel to South Africa for ARTs, recruited through the snowball sampling method; six women attending the private clinic in Maputo, including three previous reproductive travelers, for a total of nine women who had undergone ARTs in South Africa; and 14 women attending the public hospital in Maputo. This research is drawn from their discourses and narratives as well as information provided by two formally interviewed reproductive health practitioners. A total of 27 interlocutors were interviewed, four of whom were interviewed more than once. The interviews were semi-structured, normally starting with some general questions and gradually coming to focus on the subject of infertility and infertility healing. Although there was a degree of contact with some men (three in total), all of the patients that were interviewed were women, and the interviews featured a one-on-one format. This was because the women were more likely to share intimate accounts of their affliction and healing if no third parties were involved in the conversation. All of the participants provided informed consent to participate in the research.

The other research used as a basis for this reflection resulted from the preliminary results of an ongoing, anonymous, online survey, "Pregnancy, childbirth and postpartum experiences of black and African-descended mothers in Portugal", carried out by the collective SaMaNe; as of February 2022, 122 participants completed this survey. SaMaNe is a social movement founded in 2020 that aims to advance the social and academic debate concerning the various obstetric experiences of black and African-descended women in Portugal, since other groups and studies have already highlighted the experiences of obstetric violence in Portugal in the case of women in general (APDMGP 2020; Barata 2020; CEDAW 2015). Since data regarding perceptions of black women's experiences are still lacking, SaMaNe's proposal is to understand the experiences of these women with the health care system.

We have decided to collect some of our findings in this chapter to illustrate societal issues regarding the prejudiced stereotypization of black and African-descended women related to reproduction and reproductive care in Portugal and Mozambique.

3.2 Portugal and colonialism: a background

The specific colonial matrix that pertains to the regions and relations supporting the empirical examples and discussions presented in this chapter is deeply connected to a tendency toward denial – the idea of "Luso-tropicalism", which claims that Portugal was a moderate colonizer has dominated public opinion until the emergence of very recent discussions (e. g. Vala et al. 2008; Raposo et al. 2019; Maeso and Araújo 2013; Agra et al. 2021) – that is operationalized as a way to make people and the country as a whole "feel better" about a violent colonial past terminated by long and violent wars (Newitt 1995). This is a socio-historical political construction and prevails as a narrative that is supported by arguments related to soft colonization, miscegenation, and a commonly reified national entrepreneurial spirit (Cardina 2016, p. 35). This discourse tends to soften the violent aspects of colonial occupation, the enslavement of people and the slave trade since the 15th century. It also tends to soften the relationships between colonial governance and more recent political conjunctures in Portugal – in which a fascist dictatorship, the "Estado Novo", was officially in power for more than 40 years (1933–1974). These elements are crucial to understanding the symbolic and actual violence that are embedded in the exclusionary and assimilating Portuguese colonial politics, the kinds of racism that are purported by the regime, and the organic ways in which such practices today appear to be soft (Cardina 2016).

The effects of these rationales remain evident today in inequalities that are present in areas such as education, housing and justice. Several studies have managed to demonstrate that the Portuguese social, political and historical system is supported by and reproduces a racial hierarchy (Henriques 2018; Raposo et al. 2019; Roldão 2015). These remains of racial hierarchization appear to be entangled with other kinds of inclusionary and exclusionary practices based on, among other factors, constructions regarding gender and class, social networks, and bureaucracy, which can also be seen in practices and access to reproductive care.

For instance, little to no information is available regarding ethnic and cultural diversity, and no information is available with respect to access to reproductive health care by black and African-descended people in Portugal. This is mainly due to the impossibility of collecting data concerning the ethnic and racial background of citizens, since the Constitution of the Portuguese Republic (art. 13º (1)(2); art. 35º (1)(2)(3)), although it is based on a logic of non-discrimination, makes it difficult to assess existing relationships among socio-cultural and ethnic background, social class and access to services, including the ways in which structural racism "really" impacts individuals' economic welfare and their access to employment, education, and health.

3.2 Racism and reproductive bodies

Angela Davis, in her book "Women, race and class" (2016), shows how black women's bodies were exploited in the quest for greater "natural" reproduction. They became valued for their fertility, and a mother of ten or more children became a treasure (Davis 1983). These characteristics, which have been attributed to black women, have their origins in the past of slavery and colonialism and still permeate contemporary stereotyping tendencies in medical discourse and practices in reproductive medicine today (Ginsburg and Rapp 1998; Lock and Nguyen 2005; Rapp 2019; Guilfoy et al. 2008; Bhopal 2007). Biomedicine participated in these classificatory practices, having served colonization in both direct and indirect ways by producing discoveries that legitimized the management of the reproductive bodies of women in African colonial territories (Oyěwùmí 1997; Gilman 1985, 209; Comaroff and Comaroff 1992, 215; Ginsburg and Rapp 1998; Lock and Kaufert 1998).

Reproduction was indeed an important area of colonial violence and population management globally. This was the case regarding early enslaved populations, the dynamics of the slave trade and the *Partus Sequitur Ventrem* legal dispositions deployed by most colonial parties that were involved in slave trade, including Portugal (Borucki 2011; Jerónimo 2015; McManus 2020). Over time, forms of colonial population management evolved, and, during the 20th century, new dynamics started to be prioritized: after World War I, partly in response to demographic decline, colonial intervention became interested in the reproductive health of black peoples, creating policies that aimed to increase birth rates and decrease rates of sexually transmitted diseases and infant mortality. Lynn Thomas (2003), based on the case of Kenya, defines these policies as "politics of the womb", since they are specific to women and their bodies and only tangentially designed for men. These politics of the womb generally included encouraging the rapid weaning of newborns, reducing the time between births (Hunt 1998), providing tax benefits to big families, placing an increased tax burden on polygamous men and single women and, of course, the prohibition of abortion (Coghe 2020). These birth promotion campaigns also had a "civilizational" dimension, reiterating the great moral mission of the colonial project by fighting child mortality that was attributed to the "backwardness of African mothers" (Coghe 2020, p. 16; Hill 2015) who did not know how to take care of their children; it was believed that they ignored basic habits related to hygiene and child nutrition, thus preventing them from circumventing premature deaths.

Despite this reproduced racist narrative, some of these health policies met with success. However, the poor working and housing conditions, nutritional deficiencies, urban segregation, and imposed changes on gender dynamics resulting from colonization ensured that rates of malnutrition, poverty, STDs, and child mortality remained high (Coghe 2020, p. 16). Even in postcolonial configurations and under the management of independent governments in various regions, the development of better health infrastructures remains a difficulty in many sub-Saharan African countries (Dilger et al. 2012). This is partly due to the decades of direct and indirect exploitation by colonial regimes and later neoliberal development politics combined with the specific socio-political and economic configurations of each place. Although biomedical sciences have evolved beyond the scientific thinking and racial hierarchization of the colonial era, colonialism left deep wounds in the social fabric and resulted in asymmetries among social groups that are often reflected in the health indicators of these populations today (Hill 2015). Indeed, conditions of social and economic inequality combined with other environmental factors may enhance the vulnerability of populations to suffering, disease, and higher rates of mortality and morbidity (Singer et al. 2017).

4 Racism and relationships of care in reproductive health: examples from Mozambique and Portugal

4.1 Intimate lives, the options of ARTs and anxieties regarding skin color in Mozambique

There are multiple narratives produced by and around women undergoing fertility treatments. Many narratives are related to the common association between dark skin color and social inequalities, given that black people may have greater difficulty with access to education, employment and equal opportunities, thus reinforcing the stigma and the relationship between class and color (Burton et al. 2010; Henriques 2018; Roldão 2015; Taviani 2019).

This is the story of Eliana, a woman from Maputo, and part of her experience with assisted reproduction, including the doubts, anxieties and conversations it triggered. Eliana was married to Martin, whom she met during the years she lived in Europe. After a long time as a couple, both thought they had the necessary relationship, economic and career stability to have a child. Unfortunately, their attempts seemed to be in vain, as they were unable to conceive for some time – at least, for long enough that they considered resorting to ARTs.

Eliana often discussed her past anxieties regarding her inability to become pregnant and the in vitro fertilization (IVF) treatment they pursued abroad. Unlike other couples, Eliana and Martin's first IVF cycle resulted in pregnancy. However, taking into account Martin's age, Eliana told me that she had shared with a confidant her doubts and disquietudes regarding the outcome of the whole assisted reproduction process and had heard many comments and ideas that left her feeling uneasy. In these conversations, Eliana's questions mainly pertained to her partner's sperm: "What if the sperm was 'not good'? What if they had to consider using donor sperm?" At this point, issues of color emerged, with her confidant telling her that the donor should have light colored skin; Eliana heard comments about the color of the donor and was told that it would be better for the child not to be too dark. In the end, the couple managed to have a baby without having to consider the option of a donor, but this is not always the case.

Rosely Gomes Costa (2004), in a study conducted in Brazil (São Paulo) featuring medical teams and people who sought the treatment of gamete donation in public and private institutions, reported similar ideas to those discussed in the case of Eliana regarding the interference of intimate support networks. There was pressure for the couples to avoid choosing donors with dark skin; it would be rather acceptable for the child to look different from the parents if this meant that the child would be light-skinned. This choice, Costa found, was due to issues of racial discrimination that led people to conclude, as did Eliana's confidant, that a lighter child is better protected against the suffering and discrimination faced by black people.

Eliana's story alongside the findings from Brazil and Mozambique/South Africa shed light on various, more or less politically bounded and racialized, rationales regarding living expectations that seem to be present in the background of some reproductive choices. These are particularly visible when couples face the possibility of being able to interfere with the genetics of a child. The color of the gamete donor becomes relevant and triggers anxieties. These include the desire to assess genetic information and the desire to choose a lighter donor, resulting from the belief that instead of having a child that more closely resembles the parents, a lighter child has the prospect of a better life, with more fulfilled aspirations.

If ARTs raise new questions for prospective parents, on the other side of the equation, Tessa Moll (2019) explores further interconnections – those pertaining to "technologies of racialization" involved in the use and marketing of donor gametes in South Africa. Viewing this process as one of relationalities, the author highlights the importance of what she calls "curature", i.e., a "relational process managed via the powerful acts of matchers, translational figures who cull, curate, highlight, and invisibilize information during the process of matching donor gametes" (Moll 2019, p. 11). Eliana's example as well as the research of Costa and Moll highlight aspects of race-based valuation rationales that are present in discourses and imaginaries that permeate matters of kinship in the context of making both institutional and intimate choices related to people's reproductive lives.

4.2 Life at the hospital, reproduction and medical interactions in Portugal

The information available regarding reproductive health and migrant populations in Portugal is scarce and pertains mainly to migrant populations from African countries and Brazil. It shows that, overall, migrant women have worse health indicators than non-migrant women (Machado et al. 2006; Dias and Rocha 2009).

In Portugal, health professionals' stereotypes concerning migrants have been identified in episodes of discrimination. These include antipathetic care, the use of discriminatory language and hostility towards the patient (Dias and Rocha 2009, p. 138)[2]. Topa's study concerning the maternal and childcare provided to immigrant women in Porto (Topa 2015) contributes to these findings by highlighting the discriminatory attitudes of health professionals as signs of the Eurocentric view of the functioning of Portuguese institutions, which she considers to be a legacy of colonialism. Most participants in Topa's study claimed that they faced inequalities and inequities with respect to their access to health services; additionally, many explained that they pretended to ignore some unpleasant interactions to avoid retaliation from certain health professionals and institutions (Topa 2015, p. 175). This apprehension is rooted in past events and embedded in social relationships (Topa 2015). It may, however, be deconstructed by researching the manifold forms of agency and relations hidden behind it (Lock and Kaufert 1998; Benard 2016). As John Hoberman (2012) suggests, the "perceived" stoicism of black women, both in the past of slavery and colonialism and in the present, has its roots in a history of social disadvantage and continued aggression (Hoberman 2012, p. 115). Further research in this area is warranted, at least in the regions we address here, particularly concerning stereotyping practices and the power that statistical invisibility has in daily life, but also with respect to the ways in which people navigate these spaces on a daily basis.

To conduct further research to investigate these topics, the social movement SaMaNe has been collecting data concerning the experiences of pregnancy, childbirth and postpartum of black and African-descended women as part of its approach to intervening against the silencing of black mothers and women in Portugal.

Even though this research is at an early stage, its preliminary results reveal stories concerning the emergence of prejudice in biomedical health care. Below,

---

2 See also Joana Topa's research (Topa 2016; Topa et al. 2017).

we share the accounts of three women who responded to the online question-naire regarding medical experiences in pregnancy and childbirth administered by the SaMaNe collective.

The first women, who was 29 years old, stated that she had a good experi-ence during her pregnancy and childbirth, but added that: "In two moments, I felt that because I was black, they made assumptions about me; they thought I would not have a citizen card and that I already had more children. In this sec-ond moment, the doctor even suggested a hysterectomy [...]. It scared me a lot" [29 years old, Portuguese, Angolan].

In two other experiences collected, the women recalled that during labor, health professionals made assumptions about their capacity to give birth due to their racial origin: "When they injected something into my serum and the pain started, I called the nurse and said that I was not bearing the pain anymore; she told me that we, blacks, are good 'breeders' and that I would bear it" [47 years old, Portuguese, Guinea-Bissau]; "Not only did they tell my husband to go home, that he wasn't doing anything there, they also suggested that being of black descent, it would be an easy birth, and they almost never explained to me the procedures and the reason for doing them" [41 years old, Portuguese].

The situations presented are examples of common stereotyped views con-cerning black women and childbirth. These views are manifest particularly in medical or therapeutic interactions and include the tendency of clinical staff to assume particular kinds of behavior based on skin color, such as thinking that certain people will not understand the language used to communicate in the clinical setting or asking intrusive questions about peoples' private lives.

As seen previously, there are several prevalent ideas that mirror false cer-tainties regarding reproduction – "black women are child bearers *par* excel-lence", "black women are strong, more resistant to pain", "black women are more resistant to anaesthesia", "they have big breasts and large hips" – a major-ity of them without any scientific basis (Gilman 1985; Ferreira and Hamlin 2010; Benard 2016). Furthermore, stereotyped views are also implied in common pre-conceptions regarding black women's "nature" causing them to be "good" at en-during the pain of childbirth and to be prone to having many children. All of this has deprived many black women of comfort and therapeutic support during childbirth, such as by denying them pain relief. In the examples, as has been ex-plored elsewhere (Hoberman 2012, p. 115), there is a tendency for the idea of en-durance and strength to result in less attentive care from medical staff – this marks the lives of black women in many moments of their reproductive lives, and it is often considered to be a phenomenon that cannot be fought. In addition to other racist practices, these unpleasant medical interactions may deeply affect trust in health services (Moll 2021). No matter how they have been transformed, as seen in the examples, stereotypes concerning black women's hyperfertility, reproductive and physical strength, and endurance of suffering continue to prevail. Although women are by no means passive vessels for political, social and medical interventions (Lock and Kaufert 1998) and the prevalence of racist practices in health care settings cannot be generalized to encompass all settings or practitioners, there is still a long road ahead in the contexts of education, policy-making and healthcare provision if we are to transform the institutional – racialized, class-based and gender-based – regimes of invisibility that are currently perpetuated by a complex interaction of historical, socio-economic and political dimensions, which must be discussed.

5 Concluding remarks

The contextualized examples discussed above shed light on practices of racist stereotypization and further structural dimensions that contribute to their reproduction. These include the invisible aspects of statistics and the absence of information; the idea that silencing ethnic data contributes to non-discrimination, when it is actually necessary to understand its contours in specific sites; mundane comments and judgements revealing the stereotyping of black women and the attribution of specific – racialized – characteristics to their bodies; intimate conversations and what they reveal about racialized valuation rhetorics in the context of high-tech ART treatments; practices that often go unnoticed in the context of medical interactions but nevertheless shed light on the degree to which racism is still, both consciously and unconsciously, embedded in institutions; and, finally, the depiction of subalternized people as stoic and passive.

Some of these co-produced invisibilities and violences are deeply entangled with social dynamics and relationships of care. Indeed, as we have shown, in daily life, racism in biomedicine is essentially expressed in the form of negligence – such as by ignoring recurrent complaints or denying certain treatments – in treatment based on stereotypes, but it also expressed in the wider institutional absence (or non-disclosure) of data and information concerning access and care for people who belong to social and/or racialized minorities in specific territories (Moll 2021).

The conscious and unconscious embeddedness of prejudiced stereotyping rationales in clinical practice often transforms a place and moment that should be associated with protection – the quest for reproductive care and the clinical encounter – into a moment featuring the reproduction of symbolic, physical, and institutional violence that leads to suffering (Guilfoy et al. 2008; Carde 2011; Valdez and Deomampo 2019).

To believe that racism is an individual problem that can be solved only in theory removes from the debate the fact that racism consists not only of false beliefs but also of a powerful rhetorical device that was operationalized in colonial regimes as a means of legitimizing practices of domination and oppression. Despite the lack of any scientific basis for race as a biological characteristic in humans, there are rationales supported by such devices that continue to prevail, as do hierarchic considerations based on gender and class that directly and indirectly influence lives and social interactions (Russell 2015). We feel that it is not sufficient to highlight the theoretical-scientific flaws of race theory to overcome the everyday practices that reproduce colonial legacies, especially concerning the provision of health care.

We, as citizens, researchers, practitioners, and patients, must debate, discuss and promote horizontal conversations, breaking free from taboos and engaging in cross-sectorial and multidisciplinary dialogues, even if experimental, to provoke such a change.

We hope that the experiences, stories and thoughts we presented here can contribute to ongoing conversations and that more bridges and collaborations among disciplines and between science and society can be established to promote a better understanding of racism in everyday life and to support negotiations between citizens and institutions, particularly regarding invisibility and institutional racism.

References

Associação Portuguesa pelos Direitos da Mulher na Gravidez e Parto [APDMGP]. (2020): "Experiências de Parto em Portugal. Inquérito às mulheres sobre as suas experiências de parto". https://associacaogravidezeparto.pt/wp-content/uploads/2020/12/Experie% CC%82ncias-de-Parto-em-Portugal_2edicao_2015-19-1.pdf, visited on 28 January 2021.

Bagnol, Brigitte; Matebeni, Zethu; Somon, A. Blaser; Manuel, Sandra; Moutinho, Laura. (2010): "Transforming Youth Identities: Interactions Across Races/Colors/Ethnicities, Gender, Class, and Sexualities in Johannesburg, South Africa. In: *Sexuality Research and Social Policy 7*. Pp. 283–297.

Barata, Catarina. (2020): "'A mãe está calada!' O que revelam as experiências de parto das mulheres?". In *Jornal O Público*. https://www.publico.pt/2020/08/02/sociedade/noticia/mae-calada-revelam-experiencias-parto-mulheres-1925770, visited on 6 July 2022.

Benard, Akeia. (2016): "Colonizing Black Female Bodies Within Patriarchal Capitalism: Feminist and Human Rights Perspective". In: *Sexualization, Media & Society 2*. N. 4, pp. 1–11.

Bhopal, Raj. (2007): "Racism in health and health care in Europe: reality or mirage?". In: *European Journal of Public Health 17*. N. 3, pp. 238-241.

Borucki, Alex. (2011): "The Slave Trade to the Río de la Plata, 1777–1812: Trans-Imperial Networks and Atlantic Warfare". In: *Colonial Latin American Review* 20. N. 1, pp. 81-107.

Burton, Linda; Bonilla-Silva, Eduardo; Ray, Victor; Buckelew, Rose; Freeman, Elizabeth. (2010): "Critical Race Theories, Colorism, and the Decade's Research on Families of Colour". In: *Journal of Marriage and Family* 72. N. 4, pp. 440–459.

Carde, Estelle. (2011): "De l'origine à la santé, quand l'ethnique et la race croisent la classe". In: *Revue européenne des migrations internationales* 27. N. 3, pp. 31-55.

Cardina, Miguel. (2016): "Memórias amnésicas? Nação, discurso político e representações do passado colonial". In: *Configurações* 17. Pp. 31–42.

Carsten, Janet. (2000): *Cultures of relatedness: new approaches to the study of kinship*. Cambridge: Cambridge University Press.

CEDAW. (2015): *Portugal – Shadow Report for the 62nd CEDAW Session*. Lisbon.

Coghe, Samüel. (2020): "Disease Control and Public Health in Colonial Africa". T. Spears (ed.). *The Oxford Research Encyclopedia of African History*. Oxford: Oxford University Press. https://doi.org/10.1093/acrefore/9780190277734.013.620, visited on 30 September 2021.

Comaroff, Jane; Comaroff, Jack. (1992): *Ethnography and the Historical Imagination*. Boulder: Westviewpress.

Costa, Rosely. (2004): "O Que a Seleção de Doadores de Gametas Pode nos Dizer Sobre Noções de Raça". *PHYSIS:* 14. N. 2, pp. 235–255.

Davis, Angela. (1983): *Women, race, class*. New York: Vintage books.

Dias, Sónia Ferreira; Rocha, Cristianne Famer. (2009): *Saúde Sexual e Reprodutiva de Mulheres Imigrantes Africanas e Brasileiras: um estudo qualitativo*. https://www.om.acm.gov.pt/documents/58428/177157/OI_32.pdf/059d23a1-370f-49a3-a2ab-70077b24d69d, visited on 30 September 2021.

Dilger, Hansjörg; Kane, Abdulah; Langwick, Stacey [Eds]. (2012): *Transnational Medicine, Mobile Experts: Globalization, Health and Power In And Beyond Africa*. Indiana: Indiana University Press.

Fanon, Frantz. (2019 [1952]): *Black Skin, White Masks* (trans. Richard Philcox). New York: Groove Press.

Faria, Inês. (2015): "Family Re-imagined: assisted reproduction and parenthood in Mozambique". In: Kroløkke, Charlotte; Myong, Lene, Stine, Adrian, Tjørnøj-Thomsen, Tine (Eds.) *Critical Kinship Studies*. London: Rowman and Littlefield International.

Faria. Inês. (2016): "Biomedical Infertility Care and Assisted Reproduction: notes on Mozambican infertile couples' transnational therapeutic itineraries". In Duchesne, Veronique ; Bonnet, Doris. (Eds.) *Procréation Médicale et Mondialization: experiénces africaines*. Paris: L'Harmattan.

Faria, Inês. (2018): "Therapeutic Navigations and Social Networking: Mozambican women's quests for fertility". In: *Medical Anthropology: cross-cultural studies on health and illness* 37. N. 4, pp. 343–357.

Faria, Inês. (2021): "Plans, Changes, Improvisations: navigating research on the quests for fertility of Mozambican women and men". In: *Anthropology in Action* 28. N. 2, pp. 18–26.

Farmer, Paul. (2004): "An Anthropology of Structural Violence". In: *Current Anthropology* 45. N. 3, pp. 305–325.

Ferreira, Jonathas; Hamlin, Cynthia. (2010): "Mulheres, negros e outros monstros: um ensaio sobre corpos não civilizados". In: *Revista Estudos Feministas* 18. N. 3, pp. 811–836.

Gerrits, Trudie. (2012): "Biomedical infertility care in low resource countries: Barriers and Access" In: *Facts and Views in ObGyn Monograph*, N. 2, pp. 1–6.

Gilman, Sander. (1985): "How does medicine construct its objects". In: Biron J. Good (Ed.) *Medicine, Rationality and Experience*. Cambridge: Cambridge University Press, 65–87.

Ginsburg, Faye; Rapp, Rayna. (1995): *"Conceiving the New World Order: the global politics of reproduction"*. Berkeley: University of California Press.

Girard, Willliam. (2013): "'Invisibilities' Member Voices". In: *Fieldsights*, June 4. https://cu lanth.org/fieldsights/series/invisibilities, visited on 30 September 2021.

Hamed, Sarah; Thapar-Björkert, Suruchi; Bradby, Hannah; Ahlberg, Beth. (2020): "Racism in European Health Care: Structural Violence and Beyond". In: *Qualitative Health Research* 30. N. 11. https://journals.sagepub.com/doi/full/10.1177/1049732320931430, visited on 29 September 2021.

Havik, Philip. (2013): "The Colonial Encounter Revisited: anthropological and historical perspectives on brokerage". In: Cardeira da Silva, Miguel (Ed.): *The Jill Dias Lessons: antrhopology, history, Africa, academia*, pp. 97–111. Lisbon: Etnográfica.

Henriques, Joana. (2018): *Racismo no País dos Brancos Costumes*. Lisbon: Tinta da China.

Hill, Jonathan. (2005): "Beyond the Other? A postcolonial critique of the failed state thesis". In: *African identities* 3. N. 2, pp. 139-15.

Hoberman, John. (2012): *Black and Blue: the origins and consequences of medical racism*. Berkeley: University of California Press.

Jerónimo, Miguel. (2015): *The 'Civilising Mission' of Portuguese Colonialism, 1870–1930*. London: Palgrave McMillan.

Kilomba, Grada. (2008): *Plantation Memories: episodes of everyday racism*. Berlin: Verlag.

Latour, Bruno. (2010): *On the Modern Cult of the Factish Gods*. Durham: Duke University Press.

Lock, Margaret; Kaufert, Patricia. (1998): *Pragmatic Women and Body Politics*. Cambridge: Cambridge University Press.

Lock, Margaret; Nguyen, Vin Kim. (2005): *An Anthropology of Biomedicine*. Oxford: Wiley-Blackwell.

Maeso, Sílvia; Araújo, Marta. (2013): "A quadradatura do círculo: (anti) racismo, imigração e a(s) política(s) da integração em Portugal nos anos 2000". In: *Oficina CES* 407. Pp. 1–37.

Mbembe, Achille. (2001): *On the Postcolony*. Berkeley: University of California Press.

Mbembe, Achille. (2013): *Critique of Black Reason*. Durham: Duke University Press.

McManus, Stuart. (2020): "Partus Sequitur Ventrem in Theory and Practice: Slavery and Reproduction in Early Modern Portuguese Asia". In: *Gender and History* 32. N. 3, pp. 542–561.

Moll, Tessa. (2019): "Making a Match: Curating Race in South African Gamete Donation". In: *Medical Anthropology* 38. N. 7, pp. 588–602.

Moll, Tessa. (2021): "Medical Mistrust and Enduring Racism in South Africa". *Journal of Bioethical Inquiry* 18. Pp. 117–120.

Newitt, Maylin. (1995) *History of Mozambique*. Bloomington: Indiana University Press.

Oyěwùmí, Oyèrónkẹ́. (1997): *The Invention of Women: making an African sense of western gender discourses*. Minneapolis: University of Minnesota Press.

Raposo, Otávio; Alves, Ana; Varela, Pedro/Roldão, Cristina. (2019): "Negro drama. Racismo, segregação e violência policial nas periferias de Lisboa". In: *Revista Crítica de Ciências Sociais* 119. Pp. 5–28.

Rapp, Rayna. (2019): "Race & Reproduction: An Enduring Conversation". In: *Medical Anthropology: cross-cultural Studies in health and illness* 38. N. 8, pp. 725–732.

Roldão, Cristina. (2015). *Factores e Perfis de Sucesso Escolar "Inesperado". Trajectos de Contratendência de Jovens das Classes Populares e de Origem Africana.* Tese de Doutoramento em Sociologia. Lisboa: ISCTE-IUL. https://repositorio.iscte-iul.pt/bit stream/10071/9342/1/Tese%20Doutoramento_Fatores%20e%20Perfis%20de%20Sucesso %20Escolar%20Inesperado_CRoldao2015.pdf, visited 12 July 2022.

Russell, Camisha. (2015). "The Race Idea in Reproductive Technologies: Beyond Epistemic Scientism and Technological Mastery". *Journal of Bioethical Inquiry* 12. N. 4, pp. 601–612.

Singer, Merril; Bulled, Nicola; Ostrach, Bayla; Mendenhall, Emily. (2017): "Syndemics and the biosocial conception of health". In: *Lancet* 389. N. 10072, pp. 941–950.

Taviani, Elena. (2019): "Das políticas de habitação ao espaço urbano. Trajetória espacial dos Afrodescendentes na Área Metropolitana de Lisboa". In. *Cidades Comunidades e Territórios* 38. Pp. 57-78.

Thomas, Lynn. (2003): *Politics of the Womb: Women, Reproduction, and the State in Kenya.* Berkeley: University of California Press.

Topa, Joana. (2016): *Cuidados de Saúde Materno-Infantis a Imigrantes na Região do Grande Porto: percursos, discursos e práticas.* https://www.om.acm.gov.pt/documents/58428/ 179891/Tese47.pdf/e0bbb35f-ce78-4bb6-abaf-72720622af7e, visited on 20 September 2021.

Topa, Joana; Nogueira, Conceição; Neves, Sofia. (2017): "Maternal Health Services: an equal or framed territory?" In: *International Journal of Human Rights in Healthcare* 10. N. 2, doi: 10.1108/IJHRH-11–2015–0039.

Trouillot, Michel. (2003): *Global Transformations: Anthropology and the Modern World.* New York: Palgrave Macmillan.

Vala Jorge; Lopes, Diniz; Lima, Marcus. (2008): "Black Immigrants in Portugal: LusoTropicalism and Prejudice". In: *Journal of Social Issues* 64. N. 2, pp. 287–302.

Valdez, Natali; Deomampo, Daisy. (2019): "Centering Race and Racism in Reproduction". In: *Medical Anthropology: cross cultural studies on health and illness* 38. N. 7, pp. 551–559.

Judith van de Kamp
Tackling "Othering" by Reinventing International Medical Electives

Abstract: This chapter, based on extensive ethnographic research in a rural hospital in Northwest Cameroon, contributes to scholarly debates about the value of International Medical Electives (IMEs). It aims at filling the knowledge gap at hospital level on staff dynamics and its short-term and long-term effects on building equitable international relationships in healthcare. The findings show that the international students as well as the Cameroonian health workers and students engage in all kinds of processes of othering that contribute to the establishment of a dichotomy between "us" and "them". Although this is a way of coping with challenging situations, it hinders the building of stronger and equitable relationships, both at individual level (visiting student to rest of the staff, and vice versa) and institutional level (education institute to hospital and vice versa). The aim of this chapter is to encourage education institutes in High-Income Countries (HICs) to take up their responsibility to prevent harmful interactions, and improve their international internship support programs. The real-world examples can be used in pre-departure trainings to encourage students to practice reflexivity. This will also contribute to the agenda of the Sustainable Development Goals.

1 Introduction

One day, in a hospital in Cameroon, a baby was born premature and needed special care. Against the advice of the midwives and gynecologist, the mother-in-law decided that the child was to be taken back home. While the medical staff moved on attending other patients, the Belgian medical student on his internship was troubled. Convinced that the baby would die at home, he managed with some effort to convince the mother-in-law that the child should stay. On his online blog he wrote: "Okay, one life saved".

It did not take long, however, before the Belgian medical student started to second-guess his actions. He realized that, from an outsider perspective, he was unable to fully understand the situation. He worried that he had put the mother at risk of being cast out by her in-laws, now that she had gone against the initial

Judith van de Kamp, University Medical Center Utrecht

https://doi.org/10.1515/9783110765120-012

advice of her mother-in-law to take the child home. He worried that the family would flee the hospital in the middle of the night, without the child. He wrote on his blog: "Maybe I should have let events take their course and accept that this is Africa".

The above situation took place in a hospital in the Northwest province of Cameroon where I studied the interactions and power dynamics between medical staff and students from High Income Countries (HICs) and Cameroonian staff and students (van de Kamp 2017). It illustrates some of the challenges related to International Medical Electives. These challenges were often related to the international students' misunderstandings of situations at the medical workplace, which lead to all kinds of "processes of othering" as coping mechanisms, often with the effect of – intentionally or unintentionally – creating a distance between "us" and "them", hindering the building of strong relationships with health workers involved.

This chapter is about these processes of othering in the context of international medical electives (IMEs) on the level of health worker and student interaction, based on empirical findings from ethnographic research carried out between 2012 and 2014 in a rural hospital in Northwest Cameroon. It shows the expectations, perceptions and practices of both international students and Cameroonian health workers and students, both in relation to specific situations at the medical workplace and about "Africa" and "the whites" in general. The aim of sharing these insights is to stress the importance for HIC education institutes to prepare their students for IMEs by addressing othering and its effect in their student support programs. This is a crucial step towards decolonizing global health (Abimbola et al. 2021; Büyüm et al. 2020), to open a dialogue about the effects of othering on the ground. The examples of challenges, from perceptions of not only the HIC students but also the local health workers and students involved, can be used in pre-departure trainings to encourage students to practice reflexivity. This will contribute to the building of stronger and equitable relationships between HIC and LMIC institutions as well as individuals. Additionally, it contributes to the agenda of Sustainable Development Goals (SDGs), specifically the implementation of inclusive and equitable quality education (SDG4), reduced inequalities among countries (SDG10), and strengthening global partnership for the goals and for sustainable development (SDG17).

2 International medical electives

In the past 15 years, international medical electives (IMEs) in low- and middle-income countries have become increasingly popular among health students at higher education institutes in high-income countries (Drain et al. 2007; Kalbarc-zyk et al. 2019; Ravdin et al. 2006). Many students in nursing, medicine and other health sciences visit hospitals in LMICs to do internships for a number of weeks or months. Universities and other higher education institutes in HICs have welcomed this trend based on the general belief among the HIC education institutes that IMEs offer a great opportunity for students to grow knowledge, skills and attitude vital for becoming a health professional (Drain et al. 2007; Kraeker and Chandler 2013; Miller et al. 1995; Thompson et al. 2003). They have designed and put in place IME support programs – consisting of pre-depar-ture training (PDT), sometimes in combination with a post-return debriefing (PRD) – to help prepare students for their international internships and maxi-mize the benefits[1]. It is argued by various scholars that a short-term stay in a LMIC enhances communication skills and mutual respect for other cultures (Walsh 2004). It is also considered to lead to positive changes in young students' relationships and worldviews (Walling et al. 2006). Walsh (2004) argues that people who temporarily work in a setting with limited resources learn to be flex-ible and innovative.

However, much of what is known about these benefits of IMEs is solely based on HIC students' perspectives, rather than also on LMIC staff involved[2] (Velin et al. 2022). HIC education institutes study their programs by asking their students about the outcomes. For instance, at Duke University, residents were surveyed (Miller et al. 1995). One of the results was a positive effect on the personal lives of the participants, including their understanding of the world and socio-cultural factors. Researchers from the Maastricht University Medical School in The Netherlands conducted interviews with students upon their return (Niemantsverdriet et al. 2004). They identified six domains in which learning outcomes were reported: Medical knowledge, skills, international health care organization, international medical education, society and culture, and personal growth. Researchers from the United States of America conducted a literature review to capture perspectives from both American and Canadian

---

1 These programs are far from standardized in curricula, the programs' objectives and structure vary widely (Goecke et al. 2008; Huish 2012), and knowledge on the effectiveness of IME support programs is limited because evaluations are scarcely published (Kalbarczyk et al. 2019). Some of these education institutes have arranged their IME objectives around the CanMEDS framework for competencies (Dharamsi et al. 2010; Goecke et al. 2008; Purkey and Hollaar 2016; Valani, Sriharan, and Scolnik 2011). The CanMEDS competencies are: Medical Expert, Communicator, Collaborator, Manager, Health Advocate, Scholar, and Professional (Frank and Danoff 2007).
2 This is not only the case for HIC students, but also for HIC qualified workers who go on short-term missions, such as surgical teams (Velin et al. 2022).

medical students on the educational effects of their internship experiences in LMICs (Thompson et al. 2003). It was argued that IME appeared to have positive educational effects on the students' clinical skills, attitudes and values such as idealism and an interest in serving underserved populations, and knowledge of tropical medicine. The problem with such studies based on HIC students' perspectives is that they do not address relevant aspects of IMEs that the HIC students' were unable to notice, identify and put to words. This is why such studies often provide a limited view on the effects of IMEs.

Gaining in-depth knowledge on the possible negative effects of IME is highly valuable for HIC education institutes to use in their IME support program designs, with the purpose of taking full responsibility in guiding their students to avoid it (de Zeeuw et al. 2019). This knowledge can also help to open up a dialogue between the HIC sending organization and LMIC hosting organization about mutual benefits and how to create an equitable relationship between the people involved. And although there has been an increase in scholarly attention in recent years, both in the practice of IMEs and the practice of short-term health work or "voluntourism" in hospitals, for perspectives of people from the host organizations (Kraeker and Chandler 2013; Loiseau et al. 2016; McWha 2011), there has not been extensive empirical research on what happens at hospital-level when HIC visiting students work together with local hospital staff and students.

This chapter is written based on extensive ethnographic research in a rural hospital in Cameroon. The research was carried out as part of my PhD research at the University of Amsterdam, which resulted in the dissertation "Behind the smiles: Relationships and power dynamics between short-term westerners and Cameroonian health workers in a hospital in rural Cameroon" (Van de Kamp 2017). The insights used in this chapter are specifically related to international medical electives, with the aim to contribute to the development of IME support programs of HIC education institutes.

3 Methodology

The empirical data collection took place in Saint Elizabeth's Catholic General Hospital Shisong in the Northwest Province of Cameroon. The first time I set foot in the hospital was in November 2012. I received permission to conduct the research by the matron of the hospital and the *fon* of "Nso", the local chief[3].

---

3 I was also in the possession of a research permit from the *Ministere de la Recherche Scienti-*

I studied the daily interactions between the Cameroonian and visiting staff and students between February 2013 and June 2014. The methods used were semi-structured interviews, observations, informal conversations, questionnaires, focus group discussions, and the analysis of online blogs. Combining all these methods proved particularly useful, as it enabled me to identify ambiguities, for instance between what people said in interviews on the one hand, and what they did as discovered during observations on the other. I was also able to ask questions during conversations and interviews about what I had seen during observations, such as arguments between people. All data was analyzed in The Netherlands using the qualitative data analysis program NVivo. In this program, data was tagged, after which it was sorted into broad categories.

Within the 16 months of fieldwork, the hospital was visited by 66 people from high-income countries: Belgium, The Netherlands, Italy, Germany, Great Britain, Austria, the United States of America and Canada. At that time, the hospital had 315 beds, and on average 112 consultations is conducted on a daily basis, with the number of workers fluctuating around 400. Of these 66 visitors, 33 were students doing an internship in the hospital. The undergraduates were nursing students and midwifery students, and the postgraduates were medical students and residents. They either came in groups or individually.

In this chapter, I refer to Cameroonian workers as CAM workers, and to international visiting workers and students from high income countries as HIC workers and HIC students.

4 Results

The focus in this chapter is on health workers' interactions and its effects in the context of IMEs. Within the 16 months of fieldwork, it became clear from watching HIC students come and go and work together with the CAM workers, that these interactions were ambiguous. As ways to cope with situations, many workers seemed to engage in both "processes of unifying" and "processes of othering". In unifying processes, similarities between people are emphasized, such as "We are all human" and "We all want best for our patients". The emphasis

fique et de l'Innovation (MINRESI) in Yaoundé through the assistance of the International Research and Training Center, also situated in Yaoundé. A requirement for obtaining a research permit is to have the full support of a scientist affiliated to a Cameroonian university. Professor Paul Nchoji Nkwi was happy to be my local counterpart. He is Emeritus Professor in Anthropology and Deputy Vice Chancellor for Academic Affairs at the Catholic University of Cameroon (CATUC) in Bamenda.

is on what people have in common, and it enables people to connect and build a relationship. In processes of othering, it is the differences rather than the similarities that are emphasized. Whether deliberately or not, it is a way to say "you are not like me". In the context of this research, processes of othering were often collective, meaning that people would emphasize the similarities and differences between "groups". These groups were not always defined, but the terms people used were "us" and "them", distancing themselves from others by explicitly or implicitly saying "they are not like us" (Sin 2009).

Both processes of unifying and othering are coping mechanisms, used by CAM staff and HIC staff, to connect but also to do justice to one's beliefs about what is right or just. To provide an in-depth understanding of the interactions, this section addresses the interaction processes with regard to three themes: 1) Building relationships and integrating in the local community, 2) On duty: Working together in the hospital, and 3) Student differentiation: Rules and privileges.

4.1 "They think we are rich". Building relationships and integrating in the local community

Both CAM and HIC workers engaged in ambiguous practices of trying to build relationships and get closer to members of the other group, while at the same time trying to keep a distance from them. While one's own – often implicit – behavior to create a distance was not always deliberate, it was often felt by those kept at a distance. In this sub-section, examples are described to show how this led to confusion and irritation.

4.1.1 Looks, gossip and rolling eyes

Most HIC students did not find it easy to work with CAM staff in the hospital. In fact, it was often one of the most difficult issues they faced during their hospital work, especially because they did not know why CAM workers behaved towards them in the way they did. The following quote is from an interview with a Dutch nursing student:

> When there's a delivery on, they [the Cameroonian workers] want you to do all kinds of things for them. But they mumble and talk so fast that you can't understand them. So you have to ask them again, three times, and they give you that look, like: 'Ooh, here

we go, do we have to repeat it? You still don't understand?' And eventually they just go do it themselves, with an expression like [...] You can often see the irritation on their faces.

The above quote reveals that when CAM workers did not talk about their frustrations, HIC students knew about it through their "looks". Even when HIC workers had gotten used to the routine and procedures, they often felt that CAM workers kept a distance from them. For instance, HIC students felt excluded from conversations when the CAM workers talked in Lamnso' or Pidgin English, and when they asked them to translate into English they said they were often ignored or even laughed at. Some HIC students also felt they were gossiped about. The next interview quote is from another Dutch nursing student: "I really don't like the gossip. I just know it is about me, because they scornfully look at me from head to toe as they speak. It is not nice". Many HIC workers felt indignant about the Cameroonian workers' behavior towards them. They felt that they were not being given a fair chance to be part of the team.

4.1.2 Distrust about being poor and rich

What many HIC students did not seem to realize is that their own ways of building relationships with CAM workers were no less ambiguous. They tried to get closer to them in certain situations, but distanced themselves in others. In an interview with me, a British medical student explained that she never trusted CAM workers, as she always felt that she was being viewed as someone "to benefit from" because of her skin color:

> There haven't really been significant relationships really with people from the hospital. [Nurse superintendent] Mister Julius also, he took us on a hike, and I have spoken with him quite a lot. [...] But in some ways I feel like his appreciation of me is as a white person more than as me, which is something I dislike. Here I don't like to just be another White person, and people trying to make an effort with me because I am White, and interested in me because I am White. It feels a bit like: "What are you trying to get from me? How are you trying to benefit?" We've met a couple of nurses outside the hospital, but once again it doesn't feel so much like we are people to them.

Although the British student wanted to build relationships with the CAM workers, she kept her distance because she felt that the CAM workers wanted a relationship with her for the wrong reasons. She argued that being a white person and coming from a HIC, where the standard of living is significantly higher, hindered her from building relationships with them.

Many HIC workers said that their higher standard of living should not get in the way of building relationships with CAM workers in the hospital, but they nevertheless felt that it did because CAM workers considered it important, or gave it too much importance. Keeping their distance from CAM workers as a result of this realization, as the student from the quote above did, was one way of dealing with this. Another way of dealing with it, common among the HIC students, was trying to convince CAM workers that the difference is not as big as they might think. This is illustrated by an interview quote from a Dutch student:

> Some people are convinced that we're rich. Sometimes we try to explain what the situation is really like, and some people seem to understand that we're not. And then there are others that just don't get it. [...] One of the staff asked us if we could buy them cameras. So I said: "You can buy those here too, can't you?" He said: "Yes, but they're expensive around here". As if they're thinking: "So, you all have a camera and a laptop so you can go buy those things for me too".

In an attempt to show CAM workers that they are not rich, that they are in fact "poor students", some HIC students told the CAM workers that they did not pay for their flight tickets themselves, but rather they had been paid for by an organization or their parents. They did not understand that this nuance failed to make a difference in the opinion of most Cameroonian workers.

Many HIC students were in fact rather unaware of certain attitudes and behaviors that did indeed create a particular impression of wealth. This included walking around with laptops, digital cameras and smartphones. They often gave away gifts, and donated money and goods. Some HIC students plead for money from their friends and families back home, for instance to pay for patients' outstanding hospital bills, and they were often quite successful in this. They also made day trips and weekend trips, spending money on transportation and accommodation, and they bought shoes, clothing, souvenirs, and relatively expensive food items such as fish and chicken in restaurants. They did not realize that this behavior contributed to the fact that their claims of not being rich made little to no impact on most CAM workers. Instead, they stated that the CAM workers overestimated the difference – in living standards, that is – between "us" and "them".

4.1.3 Living like locals

Despite the HIC students public statements of wanting to experience another culture by working and living in it, their actual attitudes towards "living like locals" were rather more ambiguous. They wanted to be treated just like CAM students in some cases, but felt entitled to differential treatment in others. For example, when three Dutch students found out that they were expected to pay a higher registration fee than the CAM students, they insisted on "equal treatment" for all students. However, they also felt entitled to work only five days a week, while the CAM students worked six days a week.

Let us zoom in on living conditions. The hospital's guest accommodation is located inside the hospital premises, which is fenced and continuously guarded. It usually houses HIC workers only, because visiting CAM workers cannot afford the fee, or choose to stay elsewhere at a lower rate. Although some HIC workers went outside the hospital grounds when they were off duty, they spent most of their time inside the fence. They therefore had minimal exposure to the culture they claimed to be eager to learn more about and be a part of. On top of this, the hospital accommodation included a range of relatively luxurious facilities that the HIC workers were used to from back home. Their rooms were self-contained, provided with a European toilet, a gas cooking stove, cooking utensils, a double bed with double mattress, warm water, comfortable chairs and internet access. Most HIC workers were happy about their accommodation. Some said it was better than they expected, though others expected a bit more for the amount of money they paid for it. Common complaints – made to me, in private – were that the accommodation was not clean and that the internet was slow, although it was by far the fastest in the whole town and surrounding area.

HIC workers in general often did not realize that facilities such as internet and a European toilet were highly uncommon for local Cameroonian households. On top of this, because the guesthouses were connected to the hospital generator, the HIC workers were not disturbed by, and probably were often not even aware of, the regular power cuts that all locals were affected by, and that lasted for hours and sometimes days.

4.1.4 Food and local habits

HIC workers in general expressed their wish to try local food and local habits. They considered it part of the experience to get to know the environment, the people and their culture. And indeed, many tried the local foods – such as *fufu* (corn flour ball) and *jamajama* (cooked huckleberry leaves), which is eaten daily by many locals. It is a breakfast, lunch and dinner meal. However, many HIC workers tried it once, after which they concluded that eating *fufu* was "not for them". They preferred to cook themselves at the guest accommodation, instead of buying local dishes in the hospital canteen or at one of the many small restaurants in front of the hospital. A Dutch nursing student wrote on her blog: "Unfortunately I am no fan of *fufu*, I find it rather disgusting. [...] It's just a ball of dough with no taste at all".

Once a Dutch nursing student asked me where she could buy vegetables. I told her that there were many vegetable stalls right in front of the hospital gate. She replied that she did not mean those vegetables, for they did not look right. HIC workers' slight fear of local food and their wish to eat familiar things was also expressed in the food items they brought with them from home, and even more in their large quantities. One Dutch student brought 36 vacuum-packed hamburgers with her from The Netherlands. Two other Dutch students asked me whether Bamenda (which is a three-hour drive away) had a 'proper' supermarket, because they were running out of cheese, brought from home, faster than they thought.

An activity that some HIC workers were eager to do –, again: once – is killing a chicken by cutting its throat. Many of them considered this act part of the local culture. It is really only because people do not own fridges to store chicken meat in, that they buy the animal alive and slaughter it right before consumption, often at special occasions. One of the Dutch nursing students wrote on her blog:

> I really wanted to kill a chicken, so that was going to happen on Saturday afternoon. Well, I can tell you that I am a chicken murderer whoa hahahahaaa! It was really exciting, but I'm glad I did it!!! It took up to a minute, not nice, but I could not get the blunt knife through the flesh, but when the chicken was repositioned it was a piece of cake! [...] It was a cool experience, for all of us.

To conclude, the wish to live like locals is often about experiencing certain aspects of how locals live, and preferably only the pleasant – or sensational – ones. HIC workers do not expect their accommodation to be of a local standard. In fact, they evaluate their accommodation based on HIC standards. This was found by other scholars as well. McLennan (2014) argues that westerners "can be seen as "modeling" a lifestyle of cultural and material values that may be inappropriate, and which promotes modernization, or development of westernization" (p. 166). Other scholars argue that HIC workers engage in short-term trips "as long as it is not too uncomfortable" (Van Engen 2000, p. 22) and "so long as it doesn't compromise our safety and basic well-being" (Sin 2010, p. 988). Sin (2010) also argues that when HIC visitors live so closely to the locals, it accentuates the differences in their living standards. She also states:

> This runs [in] contrast to what has long been argued, that if volunteer (tourists) adopt a "serving" attitude, an equal relationship between volunteer tourists and hosts can be achieved. The question then is – is an equal relationship commonly established? And indeed, what constitutes an equal relationship? (Sin 2010, p. 988).

In the same article, Sin answers the first question raised above, by stating that "the westerners' living standard shows the superior positions of volunteer tourists" and that the practice of voluntourism is "far from achieving the supposed equal relationships" (Sin 2010, p. 989). And similarly, Madsen Camacho argues that while HIC involvement is "ideally rooted in mutuality and reciprocity between servers and served, issues of power and privilege can create an asymmetrical relationship between both" (Madsen Camacho 2004, p. 31). Based on the practices of differentiation of both the HIC and CAM workers in the hospital, I argue that there is indeed an asymmetrical relationship between the incoming visiting and permanent staff, and that it can actually perpetuate unequal power dynamics and hinder the building of relationships based on equality.

4.1.5 Open-mindedness versus prejudice

HIC workers' blogs show that they did not come to the hospital as open-minded as they said that they did. Their assumptions of "Africa" and "Africans" – that they did not mention to me, and were perhaps not even aware of – revealed that they did not feel very much like "them" at all. Most HIC students' assumptions were not based on clear images, but rather on vague ideas about "Africa" and its inhabitants in general, including ideas about poverty. These ideas seemed to influence the way in which they perceived Cameroonians, both within and outside the hospital setting. The following blog quotes from five different HIC students illustrate this:

> We went outside because the car that was to take us to Bamenda had arrived. Yes, you are reading this right; he was on time.

> Africa wouldn't be Africa it there wasn't a problem. [Context: Three Belgian students are at a bank in Bamenda to withdraw some cash, but there is a problem with the machine].

> For the first time in my life I see an African who is really in a hurry. [Context: The nurse superintendent gave a group of HIC students a guided tour. He was in a hurry].

> But the speed at which Africans change a tire was much greater than we expected. [Context: A group of western students go on a day trip, for fun. On the way, the taxi gets a flat tire].

> Remarkable really: we don't see many skinny people around. Most women have a substantial belly and of course real African buttocks! However, everybody is incredibly friendly and they all want to have a chat.

The above quotes show that HIC workers' presence and involvement in the hospital did not challenge them in their assumptions about Africa and Africans, but rather confirmed and strengthened their assumptions because they continued to think in terms of "us" versus "them". It is in line with the argument of various scholars that a short-term stay in a LMIC can actually reinforce cultural stereotypes (Guttentag 2009; McLennan 2014; Raymond and Hall 2008; Sin 2009).

4.2 "They make many basic mistakes". On duty: Working together in the hospital

Sin (2009) argues that the processes of othering that people engage in when they – deliberately or not – distance themselves ("us") from others ("them") can "potentially create rifts that hinder the building of strong personal relationships between volunteer and recipient" (Sin 2009, p. 496). I argue that biased thinking can indeed hinder the building of relationships. I also argue that it can work the other way around: negative experiences of interactions can contribute towards biased thinking. This sub-section elaborates on the mutual relationship between processes of othering and relationship building, mainly from the perspective of HIC students[4], whilst working together in the hospital. To deal with difficult situations, various coping strategies were deployed: 1) nicknaming, and 2) blaming and shaming. And a third important aspect that affected workers' interaction is 3) clashing ideas on medical practice. The way HIC students responded to a situation in which ideas on medical practice were different to what they were used to, had a profound impact on the workers' interaction. The three themes are described below.

4.2.1 Nicknaming

Three Belgian midwifery students experienced many struggles in working with one of the senior Cameroonian midwives. The Belgian students considered his way of working harsh and cruel: they considered him to not show enough sympathy for the women in labor, and to do things too fast and too rough. The girls talked about him a lot among themselves, and as a result of their shared experiences they had come up with a nickname for him: "Bruno the Terrible". They used this nickname when talking in their own language among themselves, but

---

4 While I strongly suspect that the CAM workers share their frustrations with HIC students amongst themselves in a similarly indiscriminate manner, I gained more access to the more implicit ideas and thoughts of HIC students through their online blogs.

also to me and to their friends and relatives through their online blogs. They talked about him, or rather about his caricature, in a way that made them laugh about his behavior. The exaggeration and humor seemed to help them cope with having to work with him. The following quote is from one of their blogs:

> When Wendy [one of the Belgian students] arrived at the ward, she could do a delivery of a first baby. The woman did not push very well, and because the head wasn't moving Mister Bruno had to apply fundus pressure (pressure to the top of the womb). Wendy was on her own to catch the baby. Mister Bruno, again, did his nickname 'the Terrible' proud. While he was pushing the woman's belly, he shouted at Wendy: "If you let her tear, I will shoot you". Wendy did her utmost and managed to help the woman deliver her baby without a tear (she did, however, get snipped with a blunt pair of scissors by Bruno the Terrible).

Although on the surface of things the relationship between the Belgian students and the Cameroonian midwife appeared to be fine while they worked together, the behind the scenes interactions among the Belgian students showed that it was not all fine. By nicknaming and caricaturing the Cameroonian midwife, the Belgian students engaged in a process of othering, in which there was no room for a nuanced interpretation of his behavior. The absence of a decent pair of scissors, as well as his "joke" about threatening to shoot the student if she let the woman in labor tear, were used by the Belgian midwifery students to further caricature him as "the Terrible". The effects of these processes were rather complex. On the one hand, it seemed to enable the Belgian students to better cope with the Cameroonian midwife while they worked together on the ward. On the other hand, it seemed to prevent them from establishing an open relationship with him, and creating opportunities to better understand his behavior.

4.2.2 Blaming and shaming

There were several reasons for friction between HIC and CAM students. One was that wards were often crowded with students. The ward charge and other senior nurses did not interfere much with their work and expected them to know what to do and to divide the work amongst themselves. The students thus had to "fight" one another to be able to carry out procedures that they all wanted to do. Regarding less attractive tasks such as making beds and dusting, many students preferred to wait for other students to do them. Friction and irritation between HIC and CAM students is also caused by the fact that they had different ideas about what needed to be done, who needed to do it, and most of all how it needed to be done.

I gained some insight into the HIC students' frustrations with CAM students by talking to them, though I gained a lot more insight from reading their blogs. The following quote comes from a Belgian student's blog:

> So today was the first day with the new Cameroonian students on duty. In the morning we ran into a whole group (20 teenagers) of students in the duty room [...] Our hair stood on end! And during the shift there was a lot of irritation between them and us three. The students were really pedantic right from day one. They believe they have all the answers. Incredibly frustrating when you have been working there for two months and so are fairly confident that you know more [...] Fiona got really frustrated when a second year student lectured her on how to dress a baby [...] We pointed out to her that we had already dressed plenty of babies, and by now we know exactly how it's done. To which she answered that maybe she wasn't a midwife but she was a mother. Just to clarify, this was only about the sequence in which she believed we should put the baby's clothes on, from its socks to its hat...ridiculous! We won't miss them, these students!

This quote shows how three Belgian midwifery students were irritated by the presence of CAM students. Based on her two months' work experience on the ward, the author of the blog was convinced that she knew more than them, and felt that the Cameroonian students were being pedantic and unfairly critical.

The following quote from another of the Belgian students' blogs shows that the Belgians were – without realizing it – actually doing exactly what they accused the Cameroonian students of doing:

> We were really annoyed with the students this morning. We were already making the beds and had told the other students they should get started on the vital parameters. We had to tell them at least three times that they had to get started on those. You can imagine that by the third time we weren't asking them as friendly as we did at first. They don't have an eye for the work and sometimes feel they are too good for some chores. Why the hell did you choose to work in health care if you don't want to do these things? Oh well, I shouldn't try to understand.

It did not cross the Belgian students' minds that the CAM students might have been unwilling to follow their instructions, just as they had been unwilling to follow instructions coming from the CAM students. The last sentence of the quote above furthermore reveals an unwillingness to even try to understand the behavior of the CAM students. The author seems to think of "them" as so different from "us" that she cannot imagine that she would ever be able to understand them. This example shows how HIC students' negative experiences with and ways of othering CAM workers are interconnected and they mutually reinforce one another.

Related – and most likely contributing – to HIC students' processes of oth-ering are the comments left by friends and relatives in response to their blog sto-ries. The following two comments written by relatives were in response to the blog post about the "annoying" CAM students, from which the last quote above was taken: "Don't get upset, it's not worth it. You're not going to change the mentality over there"; "Yeah, certainly not, just be patient, they'll have no choice but to learn to get their hands dirty". The HIC student then replied: "they are not going to change. They are not going to get their hands dirty; it's not in the mentality here. They sit around and wait for us to do the chores. And when something fun comes up, they are first in line, like performing deliv-eries".

This last quote shows how the HIC student's negative experiences with CAM workers contributes to and strengthens her biased thinking – and most likely that of her friends and relatives – as they explain the CAM students' negative be-havior in terms of their "mentality". This way of reasoning implies that there is no need to try to look for an explanation for their behavior, let alone relate it to one's own behavior; it is just "who they are" and "nothing can be done about it".

A final example that clearly illustrates this way of reasoning among HIC workers comes from a relative who commented on a Belgian midwifery student's blog in which the student complained about a Cameroonian student who had managed to be allowed to assist a midwife in a delivery, simply by putting on her gloves quicker than the Belgian student. The relative wrote: "Typically Afri-can!!!! But don't let them do that to you. Hug".

Simpson (2004) argues that HIC students contribute to the establishment of "a dichotomy of "them" and "us" as opposed to finding commonality between developed and developing world" (Simpson 2004, p. 688). Because of this, she wonders what these students actually learn about "the other". In line with this, Guttentag (2009) argues that the intercultural benefits of the short-term work of HIC visitors in LMICs are "possibly overstated" (Guttentag 2009, p. 546). Based on the lack of interest of HIC students in the hospital in knowing why CAM workers do things differently, and the way they engage in "processes of othering", I agree with these scholars that the intercultural competency that HIC students are assumed to gain from their international experiences – regardless of the existence of a support program of any kind – should be seriously questioned.

4.2.3 Clashing ideas about medical practice

It happened that HIC students got in a situation in which they felt they had to intervene, despite their stated intentions to learn and adapt to the local ways of working. An example comes from a Belgian medical student who joined a doctor for rounds. The doctor believed that one of his patients was suffering from malaria, and to confirm this he requested a malaria test. The Belgian student, who had been taught that more tests should be run to rule out other possible diseases and make a correct diagnosis, could not believe that the doctor was only requesting a malaria test. She had no idea that the doctor had little choice than to work this way because the patient's tight budget did not allow him to run more tests[5]. He would simply not be able to afford treatment if more tests were run. Instead of asking the doctor to explain his decision, the Belgian student tried to make him change his mind by suggesting, on the spot, that it would be better to run more tests. She did not realize that it was highly inappropriate in this hierarchical setting for a student to go against a doctor, nor was she aware of the context informing the doctor's decisions. During a one-to-one interview with me, she explained that she had felt obliged to intervene. And his reaction made her angry. She proclaimed: "He treated me like a stupid student! He simply told me I know nothing about tropical disease. I just nodded, but I know my theory well. And with children you just have to be extra careful".

In the next example, two Dutch students did engage in a dialogue with the CAM staff about why a certain practice was carried out. It was about the check-up routine for pregnant women in the hospital's primary health care clinic, where the two students worked for two weeks. However, as the students did not consider the CAM workers' answers plausible, they still decided to intervene. The clinic routine was that women came in, took a number, waited to be called for a consultation and finally received vitamins and malaria prophylaxis. The Dutch girls felt that the routine did not make sense. They believed that by changing the order of things, it would become more efficient and the women would not have to wait so long. They also assumed that no one would be against them making the change. The CAM workers involved, however, did not appreciate their intervention at all. They reacted by simply repeating their instructions more elaborately, after which the Dutch students got annoyed. The following fragment is from an interview with the two Dutch:

> Tess: They started to give a detailed explanation, like: "You have to put this here, and that there". And that got me a little annoyed, and I told them: "We are not stupid".

---

5 From the Cameroonian doctor's perspective, there is no point in knowing one hundred percent what the correct diagnosis is if there is no money left over to treat the problem. Most patients who leave the hospital without medication do not come back to collect it later. Doctors therefore adapt their medical strategy for diagnosis and treatment according to what patients can afford.

Charlie: Even though to us, the way we did it made sense. What they were doing did not make sense at all. But that's how they've grown used to doing it. The method works for them. But it really does not make sense. And we tried to explain it to them, like "Yeah, this is not very practical. Why do you do it like this?" But they couldn't explain to us why they were doing it like that.

The Dutch girls expected the CAM workers to come up with an explanation for why they were carrying out the procedure in the way they did. Because these students did not consider the CAM workers' answers plausible, they had no intention of adapting and every intention of doing it differently. Due to the Dutch girls' focus on impersonal authority, they did not consider it relevant that they were students, outsiders, and visitors for only a short period of time.

The above examples of HIC workers' interventions show how strongly they believe they are right, and often believe that they know better than the CAM workers involved. This belief serves as the legitimatization for carrying out interventions and thereby going against what many stated earlier about their intention to simply observe, adapt and learn about the workings of the hospital.

There was another case in which an HIC student decided to intervene, despite her expectation that it would not be appreciated. A Belgian medical student, who visited the hospital for six weeks, organized a meeting to improve resuscitation skills in the cardiac center after a patient had died. She told me that she knew that her plan was not going to be appreciated by the workers, but she felt she had to do it. She decided to schedule the meeting on one of her last days in the hospital, reasoning that if anyone got angry, she would at least not have to work with them for long. Another reason, I propose, might have been that she would therefore not have to know whether her intervention had been effective or not; given the likeliness of making no positive change, would it not be nicer to stay oblivious? During an interview, one HIC worker told me: "Oh well, at least we tried".

The example shows that for at least some HIC students, it is more about their efforts than the actual outcome of their efforts. This might not be enough to initiate lasting improvements, but it is enough to give HIC students the feeling that they at least tried to do something good. This finding is in line with the argument that good intentions prevail over the effect and that doing something is considered better than doing nothing and is therefore good enough (De Camp 2007; Simpson 2004). Another important remark to make here is that some HIC students know that CAM workers do not always appreciate their interventions. In addition, this does not keep them from intervening. This is (partly) attributed to the powerful belief in the impersonal authority of their superior medical knowledge, which serves as a legitimation for their interventions.

4.3 "They worship the whites". Student differentiation: Rules and privileges

This section shows that HIC students were often given privileges over Cameroonian students by the ward charges and senior nurses, who might, for instance, refrain from asking HIC workers to carry out certain tasks, such as cleaning windows and cabinets and rinsing blood out of bed sheets, which they will ask CAM students to do instead. Another example is that HIC students got away with behavior for which CAM students would be corrected or even punished, such as wearing flip flops to work, eating in the duty rooms, using mobile phones, going for breaks longer than thirty minutes and taking photographs in the hospital. Sometimes HIC students were also favored over CAM students to observe caesarian sections in the operating theater: students in the maternity ward are eligible to join a woman who needs to undergo a caesarian section, though only two are allowed at a time. This is decided upon by the ward charge of the maternity ward. Several HIC students told me that they had been appointed to go, even though there were CAM students who wanted to go as well and had been waiting for the opportunity much longer. This sub-section is firstly about the CAM students' perspectives on these practices of differentiation.

4.3.1 CAM students' perceptions and ways of coping

HIC students often had no idea about what was expected of them in the hospital in terms of rules and regulation. And even when they did know, they were often unaware of the level of inappropriateness of certain behaviors. This is explained by the example of footwear. HIC students would sometimes come to work in the wards wearing flip flops. There are three reasons why working in such footwear is inappropriate. The first is that flip-flops are locally not considered as decent footwear. Even patients and caretakers would not think of visiting the hospital wearing them. They are used to go to the hospital bathroom only. Secondly, it is against the hospital staff's house rules: all workers are required to wear black shoes with a closed front. Thirdly, the inspection would not allow it. When HIC workers would come to work in the maternity ward wearing flip flops, the ward charge often kindly asked them whether they had black shoes to wear instead. Some HIC workers did change into more appropriate shoes, while others continued to wear flip flops. The ward charge would not intervene, she would leave this. When I asked the ward charge, in private, whether she would allow CAM students or workers to wear flip flops, her answer was clear: "Absolutely not".

During two focus group discussions (FGDs) with me, the CAM students told me that "compared to us, the Whites are free". This concerned more than wearing flip-flops. They explained how HIC students could go on one-hour lunch breaks, while they themselves would have to work an extra day if they took a lunch break any longer than 30 minutes. They also explained how HIC students used their phones in the duty rooms, while the matron would take away their own phones and perhaps never give them back again if they were caught doing the same. When I asked the Cameroonian students how they felt about these differentiated practices, these were some of their replies:

> These privileges might be acceptable for people who think that white people are superior to black people. I myself [...] I think we are all the same. And if we have the same training, we should be treated the same. So I think there is a problem if they are treated differently"; "I am not comfortable. Because I always ask myself: If I go to Europe, would they treat me like they are being treated here? And I know they would not treat me the same.

For the CAM students, the privileges enjoyed by HIC students in the hospital are unacceptable, but they blame different groups of people involved. Some students blame the CAM nurses and doctors for giving HIC students these privileges, and for putting themselves in the role of inferiors. Other CAM students mainly blame the matron and the whole hospital administration for not informing the HIC students about the specific behavioral ethics in the hospital:

> If the administration would consider that all students are the same, the same laws should be respected by all. If they ask you not to come to the ward if you are not neatly dressed, you would consider to dress neatly before coming. Because you came here to learn something in the ward. So I blame the administration of the hospital for not cautioning them to do the right thing.

None of the CAM students I talked to, however, felt like standing up to the CAM nurses and doctors or the hospital administration to share their frustrations. They argued that it would only make things worse for themselves.

Another topic elaborately discussed during the FGDs with the CAM students was regarding their ideas about "why" the CAM workers differentiated between them and HIC students. They mentioned several reasons. The first was that HIC students are not used to the local setting. As one of the CAM students said:

> The visiting whites are not used to certain things, for example washing, brushing or carrying things on your head. I remember one girl saying she had never carried anything on her head. Another western student told me that they don't know how to wash dresses, since they are having some sort of machines. Life is easy there. And our ward charges know they are not used to doing all those things. So it is very difficult for them to go and ask them to do [...] That is one reason why they are treated differently.

The next FGD explanation for it:

[Francis]:   They really treat them differently. They look at whites as being superior [...] the inferiority complex is there;

[Jerome]:   Yes!

[Francis]:   That is the problem, the inferiority complex.

[Brandon]:   Exactly. The fact that we were colonized is still having an influence on our lives today. I think most of our friends from Europe can have a certain amount of privileges, just because of the color. And I think because from the beginning we were colonized we learned that the white man is superior to a black man. So that still has an impact on our lives today.

In the next FGD fragment, two students speculate about the reason why HIC students are "worshipped" by the CAM workers:

[Carine]:   The way the workers behave to the whites is like they worship the whites. I have struggled to really find out why they do so. [...] Maybe it is for personal relationship, we don't know;

[Noella]:   Yeah, because, I forgot in which ward, but I saw one time that a nurse was given a nursing watch. Maybe because of the relationship they had. Because if you discover that if you make friends [...] showing them procedures and maybe inviting them outside work, by the time the whites are leaving, they give.

The conclusion that these two CAM students reached for why HIC students are treated preferentially is because the CAM workers are trying to encourage them to give gifts.

4.3.2 HIC students' and ways of coping

The extent to which HIC students were aware of this differentiation varied. Many compared the way they were treated not with the way CAM workers were treated, but with how they were treated back home. For instance, many HIC students worked from Monday to Friday, because this is normal in their home country. They did not consider working five days in the hospital a privilege, even though CAM workers and students all worked six days a week. Many were not even aware that Cameroonians work an extra day each week. One day, I heard that the matron was irritated by the behavior of some Dutch students who were staying for three months, because they were loud and smoked a lot. It was, however, difficult for the HIC workers to know exactly how they were supposed to behave, as neither the matron nor other CAM workers and students informed them about the house rules. The HIC workers, for instance, only smoked in their accommodation, but they had no idea that it was prohibited for CAM students to smoke at all during their school years, in any location.

The fact that HIC students were not informed about the local behavioral ethics by anyone did not mean that they were unable to know: they could ask about it or simply observe what others were doing or not doing. Some HIC students were more interested than others in trying to adapt to the local ways of doing things. Some HIC students, for instance, did not eat and drink in the duty room because they did not see anyone else doing it, while others ate and drank regardless. It was often unclear to me whether HIC students did not "know" about certain rules, or whether they simply did not "want" to know. An example of this unclarity relates to the taking of photos in the hospital. All over the hospital there were signs indicating that it is prohibited to take photos. I was told by one of the doctors that the reason for this is that HIC workers in the past had posted photos of the hospital's "imperfections" online, which had upset the hospital administration. Many HIC workers, however, did bring their digital cameras to work and were eager to take photos of different aspects of their work, such as patients, wounds, colleagues and equipment. When I asked HIC students about taking photos, they replied that they had permission to do so; many had asked one of the senior nurses for permission, which had been granted. Regarding the abundant signs prohibiting photography in the hospital, they therefore either took no notice of them, or simply assumed that they were not intended for them. On top of this, they most likely did not realize that Cameroonian workers would never dare to ask for permission to do something that is obviously prohibited. HIC workers therefore did not consider being allowed to take photos a privilege.

The following two blog quotes illustrate that HIC workers were aware of certain practices of differentiation, and that although they did not understand the reasons behind them, they are nevertheless quite happy about it. The first quote is from a Belgian student's blog, and the second from the blog of a qualified Belgian midwife: "Around here, a life is not worth all that much in their eyes, but when there is a white person sitting on your bike or in your taxi you have to be extra careful as a Cameroonian. What a perk to have, they'll drive twice as careful with us and sometimes that is no luxury around here!"; "The bikers also know they have to be more careful with white people".

In addition to that, it happened that HIC students were favored over CAM students by the maternity ward charge to go and observe a caesarian section, even though CAM Cameroonian students had often been waiting for that opportunity much longer. One Dutch student whom I spoke about her chance said that she got "strange looks" from the CAM students. She was not too bothered though, because she was excited to observe a caesarian section for the first time. It remains unclear exactly how HIC workers looked at their own role with regard to these differentiation practices, and whether they realized that if and how they accepted or refused privileges had an effect on the CAM students. They might have simply argued that it were the "Cameroonians" who made the decisions, not them.

5 Discussion and conclusions

This chapter, based on extensive empirical research on what happens at hospital-level when students from HICs work together with local hospital staff and students in a hospital in Cameroon, contributes to the scholarly debates about the value of IMEs, both for the international students and the people involved at the host institutions. Although it was argued by various scholars that HIC workers gain intercultural competencies, communication skills and respect for other cultures from working in an LMIC, these findings show at the very least that this is not necessarily the case. In fact, the findings show that HIC students engage in all kinds of processes of othering that actually contribute to the establishment of a dichotomy between "us" and "them". In various situations, the HIC workers' involvement in the hospital did not challenge them in their assumptions and stereotypes, but rather confirmed and strengthened them, as a way to cope with or make sense of situations they got involved in. I argue that it is of utmost importance for HIC students to be aware of these mechanisms in the context of power and privileges, because the effect of these processes of othering is that unequal North-South relationships are being perpetuated.

There is huge potential to improve the value of IMEs. Because these electives are part of HIC higher education institutes' curricula, I argue that these institutes have the responsibility to design student support programs that adequately address power dynamics and the HIC students' role in that. A possible way to do that is by encouraging students to use reflexivity. The scholars Guilleman and Gillam (2004) argued that reflexive researchers are "better placed to be aware of ethically important moments as they arise and will have a basis for responding in a way that is likely to be ethically appropriate, even with unforeseen situations" (Guillemin and Gillam 2004, p. 277). I argue that the same can be true for students on IMEs. In addition to encouraging students to use reflexivity, I would like to draw from scholars Hunt and Godard in 2013, who – again, in the context of global health research – argued that questions of justice can help students address justice related to their own research, while at the same time supporting a process of continued international dialogue and partnership development (Hunt and Godard 2013). I argue that encouraging a discussion on justice and equity would be good to apply to international internship support programs for IMEs as well.

A reinvention of IME international internship support programs designed by HIC education institutes will lead to three crucial positive effects. Firstly, it prepares students for an elective that will indeed have great potential for enhancing intercultural competence, increasing communication skills and respecting other cultures and ways of practicing medicine. Secondly, and closely related to the first effect, the staff workers and students in the LMIC medical settings that work directly with these students benefit from it, because of the open dialogues and increased mutual understanding. This enables for many more steps towards the decolonization of global health. Lastly, it contributes to the agenda of Sustainable Development Goals (SDGs). Specifically, it contributes to the implementation of inclusive and equitable quality education, of which the IME are a part (SDG4). It also contributes to reduced inequalities among countries (SDG10). And it contributes to strengthening global partnership for the goals and for sustainable development (SDG17). All in all, it offers HIC higher education institutes the opportunity to contribute to equitable relationships, both on the individual and institutional level.

References

Abimbola, Seye; Asthana, Sumegha; Montenegro, Cristian; Guinto, Renzo; Tanko Jumbam, Desmond; Louskieter, Lance; Munge Kabubei, Kenneth; Munshi, Shehnaz; Muraya, Kui; Okumu, Fredros; Sahal, Senjuti; Saluja Deepika; Pai, Madhukar (2021): "Addressing Power Asymmetries in Global Health: Imperatives in the Wake of the COVID-19 Pandemic". In: *PLoS Medicine* 18, No. 4, pp. 1–12.

Büyüm, Ali Murad; Kenney, Cordelia; Koris, Andrea; Mkumba, Laura; Raveendran, Yadurshini (2020): "Decolonising Global Health: If Not Now, When?" In: *BMJ Global Health* 5: pp. 1–4.

De Camp, Matthew. (2007): "Scrutinizing Global Short-Term Medical Outreach". In: *Hastings Center Report* 37, No. 6, pp. 21–23.

Dharamsi, Shafik; Richards, Mikhyla; Louie, Diana; Murray, Diana; Berland, Alex; Whitfield, Michael; Scott, Ian (2010): "Enhancing Medical Students' Conceptions of the CanMEDS Health Advocate Role through International Service-Learning and Critical Reflection: A Phenomenological Study". In: *Medical Teacher* 32, No. 12, pp. 977–82.

Drain, Paul; Primack, Aron; Hunt, Dun; Fawzi, Wafaie; Holmes, King; Gardner, Pierce (2007): "Global Health in Medical Education: A Call for More Training and Opportunities". In: *Academic Medicine* 82, No. 3, pp. 226–230.

Frank, Jason; Danoff, Deborah (2007): "The CanMEDS Initiative: Implementing an Outcomes-Based Framework of Physician Competencies". In: *Medical Teacher* 29, No. 7, pp. 642–47.

Goecke, Michelle; Kanashiro, Jeanie; Kyamanywa, Patrick; Hollaar, Gwendolyn (2008): "Using CanMEDS to Guide International Health Electives: An Enriching Experience in Uganda

Defined for a Canadian Surgery Resident". In: *Canadian Journal of Surgery* 51, No 4, pp. 289–295.

Guillemin, Marilys; Gillam, Lynn. (2004): "Ethics, Reflexivity, and 'Ethically Important Moments' in Research". In: *Qualitative Inquiry* 10, N. 2, pp. 261–80.

Huish, Robert. (2012): "The Ethical Conundrum of International Health Electives in Medical Education". In: *Journal of Global Citizenship & Equity Education* 2, N. 1, pp. 1–19.

Hunt, Matthew; Godard, Beatrice (2013): "Beyond Procedural Ethics: Foregrounding Questions of Justice in Global Health Research Ethics Training for Students". In: *Academic Medicine* 82, N. 3, pp. 226–30.

Kalbarczyk, Anna; Nagourney, Emily; Martin, Nina; Chen, Victoria; Hansoti, Bhakti (2019): "Are You Ready? A Systematic Review of Pre-Departure Resources for Global Health Electives". In: *BMC Medical Education*19, N. 1, pp. 1–10.

Kamp, van de, Judith (2017): *Behind the Smiles: Relationships and Power Dynamics between Short-Term Westerners and Cameroonian Health Workers in a Hospital in Rural Cameroon.* Amsterdam: UvA DARE.

Kraeker, Christian; Chandler, Clare (2013): "'We Learn From Them, They Learn From Us': Global Health Experiences and Host Perceptions of Visiting Health Care Professionals". In: *Academic Medicine* 88, N. 4, pp. 483–87.

Loiseau, Bethina; Sibbald, Rebekah; Raman, Salem; Darren, Benedict; Loh, Lawrence; Dimaras, Helen (2016): In: *Journal of Tropical Medicine* 2569732. doi: 10.1155/2016/2569732.

McWha, Ishbel. (2011): "The Roles of, and Relationships between, Expatriates, Volunteers, and Local Development Workers". In: *Development in Practice* 21, N. 1, pp. 29–40.

Miller, William; Corey, Ralph; Lallinger, Gunther; Durack, Gunther (1995): "International Health and Internal Medicine Residency Training: The Duke University Experience". In: *The American Journal of Medicine* 99, N. 3, pp. 291–97.

Niemantsverdriet, Susan, Majoor, Gerard; Van Der Vleuten, Cees; Scherpbier, Albert (2004): "'I Found Myself to Be a down to Earth Dutch Girl': A Qualitative Study into Learning Outcomes from International Traineeships". In: *Medical Education* 38, N. 7, pp. 749–57.

Purkey, Eva; Hollaar, Gwendolyn. (2016): "Developing Consensus for Postgraduate Global Health Electives: Definitions, Pre-Departure Training and Post-Return Debriefing". In: *BMC Medical Education* 16, N. 1, pp. 1–11.

Ravdin, Jonathan, Peterson, Philip; Wing, Edward; Ibrahim, Tod; Sande, Merle (2006): "Globalization: a new dimension for academic internal medicine". In: *The American Journal of Medicine* 119, N. 9, pp. 805–810.

Simpson, Kate. (2004): "'Doing Development': The Gap Year, Volunteer-Tourists and a Popular Practice of Development". In: *Journal of International Development* 16, N. 5, pp. 681–92.

Sin, LuH Harng. (2009): "Volunteer tourism: 'Involve me and I will learn'?" In: *Annals of Tourism Research* 36, N. 3, pp. 480–501.

Thompson, Matthew; Huntington, Mark; Hunt, Dan; Pinsky, Linda; Brodie, Jonathon (2003): "Educational effects of international health electives on US and Canadian medical students and residents: a literature review". In: *Academic Medicine* 78, N. 3, pp. 342–347.

Valani, Rahim; Sriharan, Abi; Scolnik, Dennis (2011): "Integrating CanMEDS competencies into global health electives: an innovative elective program". In: *Canadian Journal of Emergency Medicine* 13, N. 1, pp. 34–39.

Velin, Lotta; Lantz, Adam; Ameh, Emmanual; Roy, Nobhojit; Jumbam, Desmond; Williams, Omolara; Elobu, Alex; Seyi-Olajide, Justina; Hagander, Lars (2022): "Systematic review of low-income and middle-income country perceptions of visiting surgical teams from high-income countries". In: *BMJ Global Health* 7, N. 4, pp. 1–13.

Walling, Sherry; Eriksson, Cynthia; Meese, Katherine; Ciovica, Antonia; Gorton, Deborah; Foy, David (20006): "Cultural identity and reentry in short-term student missionaries". In: *Journal of Psychology and Theology* 34, N. 2, pp. 153–164.

Walsh, Denise. (2004): "A Framework for Short-Term Humanitarian Health Care Projects". In: *International Nursing Review* 51, N. 1, pp. 23–26.

Zeeuw de, Janine; Van de Kamp, Judith; Browne, Joyce (2019): "Medical Schools Should Ensure and Improve Global Health Education". In: *The Lancet* 394, N. 10200, 731.

Carlo Botrugno
Migrations, Healthcare, and Racism: Striving for a "Bioethics in Action"

Abstract: Bioethics emerged in the middle of the 20th century mostly as a reaction to racist crimes such as the atrocities of Nazi medical experimentation in World War II or the outrageous Tuskegee Syphilis study in the US. Since that time, bioethics has expanded its domain and strengthened its toolbox. This contributed to the protection of patients' rights, dignity, and autonomy and enhanced their awareness of the medical treatments they undergo. However, in less than a century, discrimination and racism in healthcare, which were originally the foundation of bioethics, have been overshadowed by the tendency of bioethics scholars toward the "cutting-edge" and the "technology-driven", which has relegated the cultural, social and economic factors associated with health and healthcare to the background. In this chapter, I sketch out the steps bioethics can take to tackle the inequalities and discrimination in healthcare suffered by migrants in destination societies, at the heart of which lie xenophobic and racist attitudes, practices, and policies. I conclude by highlighting the fact that caring for migrants' health must be regarded as an "ethical imperative", which can also pave the way for a "bioethics in action".

1 Introduction

Bioethics emerged in the middle of the 20th century, mostly as a reaction to racist crimes perpetrated under the auspices of healthcare such as the atrocities of Nazi medical experimentations or the outrageous Tuskegee Syphilis study conducted in the US from 1932 to 1972. Of course, since that time, bioethics has made significant advances, expanded its domain, and strengthened its toolbox, thus contributing to the enhancement of patients' rights, dignity, autonomy and awareness of the medical treatments they undergo.

Today, bioethics is a flourishing discipline, whose development is inextricably linked to the magnitude and the relevance of current technological advancements in medicine and related fields, including recent discoveries in the field of genetics and its subfields as well as new technology-driven medical treatments. These advancements have amplified the conflicts of values and interests that

Carlo Botrugno, University of Florence

https://doi.org/10.1515/9783110765120-013

occur in the contexts of medicine and healthcare and made the task of finding an adequate balance among them increasingly difficult. The need to address these challenges increases the value of the knowledge and the tools offered by bioethics today.

Paradoxically, despite the fact that racism and discrimination inspired the foundation of bioethics, in less than a century, these topics have been overshadowed by the predominant focus of bioethics scholars on the "cutting-edge" and the "technology-driven". This has relegated the relevance of cultural, social and economic factors in health and healthcare to the background. I agree with Leigh Turner (2005; 2004) when he states that bioethics scholars have not adequately cultivated and addressed the connection between bioethics and public health, as has been done in the fields of sociology and anthropology. As a result, Turner argues, bioethics debates seem to be dominated by "middle-class preoccupations and fears" (Turner 2005).

A similar concern was expressed by the philosopher and physician Giovanni Berlinguer, who advocated an "everyday bioethics" to counterbalance the dominance of "frontier bioethics" (Berlinguer 2000; 2003). The Italian National Committee for Bioethics (Comitato Nazionale per la Bioetica 2010) has referred to this distinction in the following terms:

> [f]rontier bioethics focuses on the most problematic and controversial matters involving public policies and personal choices, especially on what concerns classic boundaries (i.e., birth and death); the troublesome character of these matters often depends on this field being radically new, resulting from the continuous development of biomedical sciences as well as from new technological advancements. Conversely, everyday bioethics originates from a dimension that is much closer to individuals' common experience; rather than the exceptionality of extreme cases, it instead focuses on situations of normalcy[1] (Comitato Nazionale per la Bioetica 2010, p. 5).

If bioethics aims at "weighing the difference" (Botrugno 2018), i.e., considering the impact of cultural, social and economic factors – such as employment, income, education, gender, and discrimination – on health and healthcare access, it must strengthen its connections with social justice. Although it may seem that social justice is unrelated to health and healthcare, inequalities and discrimination indeed have a significant influence on both, especially when referring to the most vulnerable groups, such as undocumented migrants, refugees and asylumseekers (Botrugno 2014a). Since the 2000s, the analysis of these factors has assumed a renewed relevance due to the work of the WHO Commission on the Social Determinants of Health (WHO 1998; WHO 2003). Additionally, empirical re-

---

1 Translated from the original Italian.

search conducted over the past two decades to investigate the correlations between socio-economic conditions and health inequalities has shown that each of these factors has its own impact on individuals' health conditions and that each of them often negatively affects health protection, being largely responsible for a social gradient in health and healthcare access (e. g. Moor et al. 2017; Marmot 2017; European Commission 2013; Tognetti Bordogna 2013;; Reyneri 2011; EU-OSHA 2007; Lindert et al. 2009; Fazel et al. 2005; WHO 2003). This body of evidence has made it possible to show that migrant status in industrialized societies represents a cause of multi-dimensional vulnerability, the seriousness of which can undermine health protection more than any other factor (Fundamental Rights Agency 2017; 2011; Botrugno 2014a).

In this chapter, I draw from my previous work in the field (Botrugno 2014a; 2014b; 2018; 2019a; 2019b; 2020a; 2020b) to sketch out the steps that bioethics can take to tackle the inequalities and discrimination in healthcare suffered by migrants in destination societies. As I subsequently argue, at the heart of these phenomena lie xenophobic and racist attitudes, practices and policies. Although I mostly refer to the situation of undocumented migrants and asylum seekers in the European Union (EU) and the Mediterranean Sea area, many of the considerations discussed here apply to the whole spectrum of the phenomenon of international migrations.

I start by outlining the main threats to the health of migrants throughout their illegal journeys as well as the main obstacles that typically prevent them from accessing adequate healthcare services in destination countries (Section 2). Thereafter, I draw attention to the impact of COVID-19 pandemic on migrant populations (Section 3), particularly in the light of the growing pushes toward the digitalization of healthcare delivery and its potential to create further barriers to service accessibility. I then shed light on (and reject) the association between migrations and the spread of infectious diseases (Section 4), which is often used to increase political tensions regarding the management of international migration flows and to implement racist policies that prevent migrants from having full access to healthcare in destination countries. In such a context, I illustrate the shameful case of the commercial ferries used by the Italian Government to physically isolate and contain both newly arrived undocumented migrants and migrants hosted in detention centers who have tested positive for COVID-19 (Section 4.1). Subsequently, I focus on the interactions among detention, labor exploitation and healthcare, in which context migrants often remained entrapped because of their intrinsically precarious socio-economic conditions and their legal status. Thereafter, I highlight the main challenges associated with socio-cultural misconceptions in health and healthcare (Section 6) that exacerbate health inequalities to the detriment of migrant people, condemning them to "invisibility" (Section 7). In the conclusion (Section 8), I explain why caring for migrants' health must be regarded as an "ethical imperative" and thus sketch out possible ways for bioethics to contribute to this task, which also entails fostering a "bioethics in action".

2 What are the health risks faced by the migrant population?

In 2022, international migrants were estimated to number 281 million worldwide, corresponding to 3.5% of the global population (IOM 2022), a figure which is increasing at a constant pace –today, it is 41% higher than in the year 2000 (IOM 2020). This shows that the utopian dream of a world without borders that underlies the process of globalization is being abandoned in favor of a strengthening of border control and the ensuing exacerbation of political tensions and national ambitions, which is linked with xenophobia and racism (Botrugno 2014b).

A migrant is defined as an individual who is in a country other than the one in which he or she was born. In this sense, the term "migrant" is used to encompass any kind of human movement. However, when defining migration, I carefully avoid any legal provisions or statements associated with policy-makers. Instead, I frame the migration act as a "process", in which context I refer to the work of Abdelmalek Sayad (1933–1998), a French-Algerian anthropologist who applied the perspective of social phenomena as "total social facts" (Mauss 1966) to migration issues. Following Sayad (2006), migration must therefore be viewed as an "epistemological itinerary" because its analysis involves the whole spectrum of the social sciences: law, economy, history, geography, demography, sociology, anthropology, linguistics, socio-linguistics, etc. Despite the great appeal of Sayad's thought, the epistemological complexity that should inspire the analysis of migration issues, particularly its impact on people's health, has always been disregarded at political level. Conversely, dichotomizations and simplifications have been used by policy-makers and mass media to legitimize the distinction between "economic migrants", i.e. those who want to cross a border to "make money", and "asylum-seekers", i.e. those who "move our heart" because they want to escape from war and persecutions. It is difficult to believe that this rhetorical stance has become a central point of the "flows' management strategy" pursued by the EU in recent decades (Botrugno 2014b). Indeed, since preventing asylum-seekers from entry is explicitly forbidden by the 1951 UN Convention Relating to the Status of Refugees – see the so-called *non refoulement* principle (UN 1951, art. 33) – the distinction between asylum-seekers and economic migrants has enabled national authorities of EU member states to evade their duty to protect foreign individuals who are intercepted at their borders or in international waters.

Efforts on the part of the EU to implement a migration flow management strategy have been mostly directed at promoting the involvement of third countries' authorities in strengthening border controls. This includes a demand for the physical containment of undocumented migrants and asylum-seekers who are intercepted on third countries' soil or waters (Rigo 2007; Botrugno 2014b). As has been argued in this regard, the EU Member States have adopted

[a] set of seemingly disparate developments concerning the constant reinforcement of border controls, tightening of conditions of entry, expanding capacities for detention and deportation, and the proliferation of criminal sanctions for migration offences, accompanied by an anxiety on the part of the press, public and political establishment regarding migrant criminality can now be seen to form a definitive shift in the European Union towards the so-called 'criminalisation of migration' (Parkin 2013, p. 1)

As a main corollary of adopting policies that limit the legal entry of migrants, illegal migration routes have developed further, and criminal gangs leveraging migrants' desperation have flourished. As mentioned previously, over time, this has contributed to a dramatic increase in the number of deaths both in the Mediterranean Sea as well as on the way to embarkation points in third countries (IOM 2017).

Leaving the political perspective to the advantage of Sayad's epistemological itinerary therefore allows us to acknowledge that the majority of migration flows today – not only the illegal ones – can be framed as part of a much wider category, that of "displaced people" or people who are "forced" to leave their countries, often for overlapping reasons. A forced migrant is therefore someone who has little or no choice when it comes to deciding to leave his or her country, family members, personal belongings, history, culture, traditions, customs, habits, etc. As Sayad himself acknowledged, this have a dramatic impact on migrants' future wellbeing as it leads to a suffering that can be understood in terms of a "double absence" (Sayad 1999).

On this account, when analyzing the repercussions of migration on an individual's health, it is necessary to take into account not only events and conditions that are strictly related to the journey but also any other events that precede the journey and lead to its start. This is particularly relevant when referring to the criminal gangs of people smugglers and human traffickers[2] on which mi-

---

2 Although the distinction between smuggling and trafficking is very subtle, it may be useful to

grants rely in their attempts to reach EU soil. For instance, consider the case of migrants from the region of sub-Saharan Africa, who are forced to pay considerable sums of money to smugglers and to complete long and dangerous journeys before reaching the embarkation points on the Mediterranean coast of Africa. As reported by many NGOs operating in the Mediterranean Sea area (FTDES 2021; Mediterranea 2022; Border Violence Monitoring Network 2022), in addition to the psychological distress and the physical suffering related to travelling in very precarious conditions, smugglers often leverage migrants' condition of subjection to abuse them and rob them of their savings. Additionally, migrants are frequently subjected to violence, rape, and torture. Furthermore, they may be sold, executed, or abandoned in the middle of the desert or in the open sea. According to data collected by the IOM "Missing Migrants Project", 2961 migrants were found dead in 2017 alone (IOM 2017). It is therefore undeniable that illegal migration routes expose migrants' health and lives to a countless series of dangers, including injuries, hypothermia, burns, gastrointestinal illnesses, cardiovascular events, pregnancy complications, diabetes, hypertension, and psychological distress.

Once migrants have survived the smugglers' violence and the dangerous journeys, they must face a series of issues related to their adaptation to and integration into destination countries. Before the COVID-19 outbreak, national governments of EU countries such as Italy, Spain, Greece and Malta were already emphatically protesting against their (assumed) excessive exposure to migration flows from the Mediterranean Sea. This led them to petition the EU to ensure the "fair relocation" of migrants, i.e. a form of coactive redistribution of the "migration burden" to other member states. These tensions were heightened on purpose by some right-wing political parties that leveraged the fear of an "invasion of strangers" – in particular Muslim people – to exacerbate nationalist and xenophobic attitudes among native populations.

---

recall the UN Protocols on the smuggling and trafficking of people, both adopted in 2000. The "Protocol Against the Smuggling of Migrants by Land, Sea and Air" defines smuggling as "the procurement, in order to obtain, directly or indirectly, a financial or other material benefit, of the illegal entry of a person into a State Party of which the person is not a national or a permanent resident" (UN 2000a. art. 3). Meanwhile, the "Protocol to Prevent, Suppress and Punish Trafficking in Persons, especially Women and Children" defines trafficking as "the recruitment, transportation, transfer, harbouring or receipt of persons, by means of the threat or use of force or other forms of coercion, of abduction, of fraud, of deception, of the abuse of power or of a position of vulnerability or of the giving or receiving of payments or benefits to achieve the consent of a person having control over another person, for the purpose of exploitation" (UN 2000b, art. 3).

Leaving aside these attempts to promote migrants' relocation – which more closely resembles a deportation – it must be highlighted that upon their arrival in the EU, undocumented migrants face two main difficulties: getting a job and achieving a legal status. These difficulties are tightly correlated since, over time, both EU policies and immigration legislations in member states have mostly associated foreign citizens' right to stay with their ability to obtain a regular job. However, the latter is virtually impossible in practice for people who lack a legal status. This represents a problem not only for undocumented migrants but also for "overstayers", i.e. foreign nationals who entered the EU with a temporary visa – for work, tourism, or study – and remained after the period for which they were permitted residence. Despite the common perception that migrants are invading the EU via illegal disembarkation, it has been argued that the overstayers represent a significant portion of undocumented immigrants in the EU (Cuttitta 2007). Given the difficulties of obtaining a regular job without a legal status, many migrants are therefore forced to accept illegal, precarious, dangerous, and underpaid jobs on the black labor market (on this point, see also Section 5).

The overall state of subjection experienced by migrants in destination countries has a direct incidence on their health by increasing their exposure to psychosocial disorders, reproductive problems, infant mortality, nutritional disorders, noncommunicable diseases, and drug and alcohol abuse (WHO 2017a; WHO 2017b). Furthermore, the condition of illegality into which many migrants are forced prevents them from having adequate access to healthcare services, with the exception of emergency treatments. The persistent economic crisis in the EU area, alongside increasing feelings of intolerance towards foreign people, have inspired recurring attempts to introduce "zero tolerance" policies for matters pertaining to undocumented migrants' access to healthcare (Botrugno 2018; Worthing et al. 2021). In this regard, it is worth remembering that in 2009, Italy introduced the crime of "illegal stay"[3], which included an obligation for healthcare professionals to report any illegal migrants receiving care. Yet even before the Constitutional Court dismantled this provision, the obligation to report migrants was largely disregarded and strongly contested by Italian healthcare professionals as well as by many NGOs that promoted "I do not report" campaigns. Similarly, in 2012, Spain excluded undocumented migrants from receiving healthcare services with the exception of emergency treatments[4]. Even in this case, such a provision has been openly contested due to its inhumane effects,

---

3 Italian Law no. 94/2009.
4 Spanish Royal Decree no. 16/2012.

and several regional governments prevented its application and provided health-care to migrants regardless of their status. Nevertheless, the introduction of this provision has had a dramatic impact over time, with approximately 1 million health books withdrawn and thousands of migrants excluded from healthcare services (Red Acoge 2015; Reder 2017). More recently, in the UK, due to the approval of the "Charges for Overseas Visitors Regulations" (2015 and 2017), patients "not ordinarily resident" in the country have started to be charged for secondary care at 150 % of its cost (Worthing et al. 2021). In case of treatments that cannot be withheld (i. e., treatments that are urgent or immediately necessary), patients are charged retrospectively. Even in this case, the adoption of these racist provisions has given rise to indignation and calls for a reaction, ultimately leading to initiatives such as the "Patients Not Passports" toolkit[5]. However, regardless of whether these provisions are successfully implemented in practice, they are certainly capable of creating "hostile environments" (Worthing et al. 2021) that can serve as significant deterrents for undocumented migrants and other migrants in precarious living conditions seeking access to healthcare services.

3 COVID-19, migration and (digital) healthcare inequalities

In response to the health emergency provoked by the outbreak of the virus, national governments worldwide have implemented heavy restrictions, such as social distancing, quarantine, and the temporary closure of all "non-essential" economic activities (i.e. so-called lockdown measures). The severe pressure faced by healthcare systems has also led to a reduction in hospital admissions, which have been limited to patients with acute symptoms. In parallel, most countries gravely affected by the pandemic have restricted access to emergency departments and primary care services to prevent these places from becoming infection hubs (Garattini et al. 2020; Botrugno 2020a).

It is known that the availability of formal and informal resources is key to managing and reacting to negative events as well as preventing adverse outcomes (Bourdieu 1977; 1986). This also holds true in the field of health care (Kawachi 1999). In a pandemic context, an individual's capacity to control factors

---

5 This initiative advocates for people facing charges for NHS care and for taking action to end immigration checks and upfront charging in the NHS. See the "Patients Not Passports" toolkit at www.patientsnotpassports.co.uk.

related to infection exposure and to adopt mitigation strategies proves to be fundamental for remaining healthy. It has been widely acknowledged that vulnerable populations have been disproportionately exposed to the impact of the pandemic (Jaljaa et al. 2022; Gama et al. 2022; Mengesha et al. 2022; Berardi et al. 2021; Crouzet et al. 2021; Gruer et al 2021; Knights et al. 2021; Mangrio et al. 2021; Greenaway et al. 2020). These populations include detainees, homeless people, people with disabilities, elderly individuals housed in residences and undocumented migrants. During the health emergency, these population groups were exposed to more physical and psychological harms than any other (Bernardini 2020; Botrugno 2020b). This was due to the significant – and, in some cases, extreme – compression of their levels of autonomy and agency. In some cases, this even made it difficult to observe the basic norms against infection risk, such as social distancing, sanitizing hands frequently, and using personal protective equipment.

The significant limitations of fundamental rights and individual freedoms were often paralleled by multiple attempts to enhance a "digital transition" of our professional and social relationships. These attempts also involved healthcare, as multiple strategies were adopted to encourage the use of digital healthcare tools (Kaplan 2022; 2020; Botrugno 2022; Hollander and Brendan 2020). In this regard, however, it must be remembered that material access to the internet and related technologies continues to represent a critical factor in many areas in the world (Van Deursen 2020) and is therefore considered to constitute a "primary digital divide" (EPHA 2014; Latulippe et al. 2017). This must be coupled with the pre-pandemic awareness that the benefits of digital healthcare services are limited to certain populations, namely those "who already possess a much broader set of 'health skills' – including awareness, attention, ambition and self-discipline – to use new technologies for better health outcomes". These skills are the outcome of formal education, and they refer to "cognitive and behavioural habits learnt and adapted from peers in particular social contexts from an early age" (EPHA 2014, p. 16).

The use of ICT-mediated healthcare services indeed leverages existing abilities, in particular levels of health and digital literacy, thus making these services particularly attractive to people who are well-educated and can rely on personal resources, but its use remains difficult for any others (Feng and Xie 2015). This led scholars to describe "knowledge" and any knowledge-related abilities in terms of a "secondary digital divide" (McAuley et al. 2014; Kontos et al. 2012).

COVID-19 put the interplay between social and health inequalities and digital divides into sharper relief, strengthening the gap between people who are already acquainted with the functioning of the healthcare systems and people who suffer the weight of inequalities (Valeriani et al. 2022; Paccoud et al. 2021). Given that socially and economically disadvantaged people are at higher risk of suffering "from chronic health conditions and faces barriers to access health systems", digital health is "likely contributing to this unequal distribution of vulnerability. As the use of technology massively increases during the COVID-19 crisis, so do the impacts of digital inequalities" (Beaunoyer et al. 2020, 3). Some studies have established a direct connection between digital inequalities and the risk of becoming infected by COVID-19, showing that people "vary in terms of what we call their COVID-19 exposure risk profile (CERPs). CERPs hinge on preexisting forms of social differentiation such as socioeconomic status, as individuals with more economic resources at their disposal can better insulate themselves from exposure risk" (Robinson et al. 2020, p. 1). In parallel to socioeconomic status, another factor that affects this indicator is the weight of digital divides. People able to "more effectively digitize key parts of their lives enjoy better CERPs than individuals who cannot digitize these life realms" (Robinson et al. 2020, p. 1)

Digital inequalities therefore caused certain population groups to be more vulnerable to both the risk of infection and the social, psychological, and economic repercussions of the health emergency. Considering the prolonged and repeated interdictions of social contact alongside quarantine and lockdown restrictions, it is clear that internet access became pivotal for gaining access to COVID-19-related health information. Internet access was fundamental not only to becoming acquainted with restrictions and behavioral norms but also to learning the relevant guidelines and recommendations. As has been emphasized, "[w]hen individuals understand the need and rationale behind government-enforced measures, they are more motivated to comply and even adopt measures voluntarily" (van Deursen 2020, p. e1). Moreover, access to internet tools "enables individuals to share news and experiences with people they cannot meet face-to-face, remain in contact with friends and family, seek support, and ask questions of official agencies, including health agencies" (van Deursen 2020, p. e1)

ICT-mediated healthcare services also proved to be a fundamental tool for providing psychological support to mental health patients as well as for supporting other people in coping with the multi-dimensional impacts of the health emergency. Therefore, the absence of internet access and related technologies as well as the skills required to use them properly became an additional factor associated with vulnerability: "[w]ith health systems already experiencing difficulties to adequately answer the burden of mental health disorders, social distancing measures increase the weight of technology to pursue psychological therapeutic services" (Beaunoyer et al. 2020, p. 4).

This confirms that the analysis of the epidemiological distribution of COVID-19 must be framed in the knowledge domain of the social determinants of health:

> [g]lobally, it is already clear that low-socioeconomic status (SES) populations are becoming infected and dying at much higher rates than their privileged counterparts. Due to long-standing social inequalities, their risks are higher, and their communities are suffering disproportionate losses in terms of infection, death, and economic devastation due to the pandemic. Low-SES groups are also much more likely to labor in high-contact, public-facing jobs such as supermarkets; provide essential transportation services; and do essential work in congregate workplaces such as food-processing facilities (Robinson et al. 2020, p. 2).

As noted above (Section 1), racism and discrimination must (also) be considered as social determinants of health, given that they are responsible for a social gradient in health and healthcare access. The pandemic indeed showed that digital inequalities contributed to a racially oriented epidemiological situation, with black and Latino people being disproportionately affected by the impact of the virus, particularly in the US and the UK (Oppel et al. 2020; Greenaway et al. 2020). In this regard, it has been highlighted that

> as of June 2020, African Americans and Latinos have accounted for 21.8 and 33.8% of COVID-19 cases, respectively, but only make up 13 and 18% of the population. In New York City, both African Americans and Latinos were twice as likely to die from COVID-19 compared with the white population even after age adjustment. Similarly, in the UK, Black and Asian minorities were more likely to die from COVID-19 compared with those of white ethnicity (hazard ratios of 1.7 and 1.6, respectively) even after adjusting for age, underlying medical co-morbidities and levels of deprivation (Greenaway et al. 2020, p. 1).

Pandemic experience therefore adds to the available evidence suggesting that discrimination, racism, structural violence, and stigmatization lead some population groups to be deprived of healthcare. However, it also shows that digital healthcare can exacerbate these inequalities, increase the gaps between population groups, and therefore reinforce the exclusion of the most vulnerable people. Digital exclusion must therefore be viewed as an emerging form of social exclusion given the dramatic repercussions it can have on other determinants of health. Such an exclusion can indeed foster a "digital vicious cycle" (Beaunoyer et al. 2020) that contributes to worsening the resources and social deprivation of some population groups.

4 Do migrants represent a threat to public health?

The pandemic has reinforced rhetorical stances pertaining to the association between human movements and the risk of infectious disease (Berardi et al. 2021), which is a recurring topic in debates and the ensuing political tensions related to the management of international migration flows (Markel and Stern 2002). Even before the outbreak of COVID-19, the assumed association between migration flows and the spread of infectious diseases – especially HIV – was used at political level, not only in the EU, to legitimize orientations and politics directed to strengthening borders control and preventing the entry of undocumented migrants[6]. As emphasized by Mukumbang, "[t]here is a long-established pattern of linking minorities, racial groups, and specific communities to disease. In particular, immigrants have been stigmatized as the origin of a wide variety of physical and societal ills including diseases" (Mukumbang 2021, p. 3).

But do migrants' health conditions represent a threat to public health in destination countries? On paper, it seems undeniable that infectious diseases may endanger native population health, as shown by the spread of COVID-19 or, even before its outbreak, by that of syndromes such as the severe acute respiratory syndrome (SARS), the Middle East respiratory syndrome (MERS), and, more recently, West African Ebola. However, the relevance of these threats has always been controversial and has been deemed to be highly variable, i.e. depending on the specific features of each syndrome as well as on the conditions that fostered their spread (Mukumbang 2021). Nevertheless, mass media and nationalist parties often leverage the fears of the native population and use migrants as political scapegoat. Indeed, infectious diseases are not at all a peculiarity of African, Asian or South-American countries, nor do they have any relation with genetic or "ethnic" factors. Rather, they are caused by extremely poor hygienic and living conditions. Industrialized countries have significantly reduced the incidence of infectious diseases among the native population as a result of good living conditions, adequate sanitation, efficient healthcare systems, vaccination programs and antibiotics (WHO 2017a). Moreover, a large body of literature currently rejects the association between migration flows and the importation of infectious

---

6 Risk of HIV infection has been a fundamental component of the anti-immigrant rhetoric of Donald Trump both before and after the 2016 US presidential election, especially with regard to Mexican migrants. In the EU, similar arguments have been used repeatedly by political parties such as the Polish *Prawo i Sprawiedliwość*, the Italian *Lega Nord*, and the UK's Independence Party during the 2016 Brexit campaign.

diseases, which has been deemed as "epidemiologically unfounded" (e. g. National Committee for the Bioethics 2017; Castelli and Sulis, 2017; Pfortmueller et al. 2016; Arnold et al. 2015; ISS 2015). In spite of the common perception that exaggerates the risk of infection upon migrants' arrival, available evidence shows that infectious diseases among migrants have a "negligible impact" (Castelli and Sulis 2017) on the epidemiology of destination countries: "infectious diseases are not at all a health priority at hotspots and first arrival sites, where traumatic, obstetrical and psychological disorders are most prevalent" (Castelli and Sulis 2017, p. 4). In this regard, the Italian National Health Institute (ISS 2015) has reported that infectious diseases detected in groups of undocumented migrants upon their disembarkation in Italy – which, as is known, is one of the countries most frequently exposed to migration flows from the Mediterranean Sea – are mostly limited to dermatological infections like scabies and other controllable diseases, such as measles and varicella. Furthermore, the association between migration and the importation of infectious diseases has been explicitly rejected by the WHO: '[c]ommunicable diseases are associated primarily with poverty. Migrants often come from communities affected by war, conflict or economic crisis and undertake long, exhausting journeys that increase their risks for diseases" (WHO 2017a). Regarding the importation of exotic and rare infectious diseases into Europe – such as the Ebola, Marburg and Lassa viruses or Middle East respiratory syndrome (MERS) – the WHO estimates this risk as extremely low: "[e]xperience has shown that, when importation occurs, it involves regular travelers, tourists or health care workers rather than refugees or migrants" (WHO 2017a).

4.1 The (shameful) case of ferry quarantine in Italy

As discussed above, the outbreak of COVID-19 forced national governments and health authorities worldwide to adopt significant restrictions to fundamental rights and individual freedoms for the purpose of protecting population health. In this regard, the UN Office of the High Commissioner for Human Rights (OHCHR) remarked that the health emergency cannot be used as "a cover for repressive action under the guise of protecting health nor should it be used to silence the work of human rights defenders" (OHCHR 2020a). In parallel, the Human Rights Committee at the International Covenant on Civil and Political Rights reminded that, even during the pandemic, national authorities must "treat all persons, including persons deprived of their liberty, with humanity and respect for their human dignity, and they must pay special attention to the adequacy of health conditions and health services in places of incarceration"

(OHCHR 2020b). Similarly, the European Committee for the Prevention of Torture and Inhuman or Degrading Treatment or Punishment (CPT) also emphasized that the protective measures adopted to fight the pandemic "must never result in inhuman or degrading treatment of persons deprived of their liberty" (CPT 2020). This entails that, even in an emergency situation such as a pandemic, any restrictions on fundamental rights must be strictly necessary, i.e. they should be unavoidable and proportionate to the aim of protecting population health (Lebret 2020). That fact notwithstanding, as emphasized by Thomson and Ip (2020), the pandemic has "sparked authoritarian political behavior worldwide, not merely in regimes already considered to be disciplinarian or tyrannical but also in well-established liberal democracies with robust constitutional protections of fundamental rights" (Thomson and Ip 2020). In some contexts, the need to protect population health has legitimized a rationale of "permanent emergency" that represented the perfect ground for large-scale and unjustified human rights violations. These violations certainly include the case of the "ferry quarantine", i.e. the use of commercial boats by the Italian government throughout the health emergency to forcibly isolate undocumented migrants and asylum-seekers from the rest of the population. On April 7, 2020, an interministerial decree[7] endorsed by the Italian Ministry of Health declared that it was unsafe for the ports of Italy to receive migrants and asylum seekers who were rescued from ships flying foreign flags beyond the Italian Search and Rescue Area. On April 12, 2020, another decree[8] from the Protezione Civile (an emergency body depending on the Italian Interior Ministry) established that newly arrived undocumented migrants and asylum seekers should be placed into quarantine on commercial ferries rented by the Government for this purpose to mitigate the risk of infection from COVID-19 to the detriment of the native population. In January 2021, this measure was extended to undocumented migrants and asylum seekers already housed in detention centers on Italian soil who were found to be COVID-19 positive. As noted by several NGO activists (MeltingPot 2021; ASGI 2021; Buffa et al. 2021), most of these individuals were forcibly and collectively transferred without any notice, even in the middle of the night. Migrants who were released after the quarantine period have declared that the average length of the detention on the ferries varied from two to three weeks, and no lawyers, state officers or policemen were admitted on board (Gianguzza and Karkouri 2021; MeltingPot 2021; ASGI 2021). The Italian Red Cross (CRI) was

---

7 Decree n. 150 from April 7, 2020, adopted by the Italian Ministry of Infrastructure and Transportation, the Ministry of Foreign Affairs and the International Cooperation, the Ministry of Interior, and the Ministry of Health.
8 Decree of Head of Department of *Protezione Civile*, n. 1287 of 12 April 2020.

forced to take on the role of providing health and social assistance to detainees on board. This included psychological care, linguistic and cultural mediation, and care for pregnant women and vulnerable people. Nevertheless, it has been emphasized that the CRI was barely able to manage the people who were detained on the ferries, and health and living conditions were reported to be awful: "many people, including some with health problems, were forced to shout, to protest and to act violently just to get the attention of the personnel. People were shouting, beating on doors, going on hunger strikes or harming themselves just to get to talk with someone, which rarely occurred"[9] (Gianguzza and Karkouri 2021, p. 125).

According to the information available, three young men lost their lives due to the detention on the ferries. Two of them died as a result of health complications unrelated to COVID-19. Another man threw himself into the sea in an attempt to escape and reach land. From April to November 2020, at least 10,000 people were detained on the ferries and at least six ships were rented for this purpose by the Italian Government (MeltingPot 2021; ASGI 2021), which announced its intention to rent four more ferries in May 2021[10]. In May 2022, the Government approved the termination of this measure, many weeks following the cessation of the "state of emergency"[11], which occurred on March 31, 2022.

The case of the ferry quarantine has caused indignation, particularly from human rights advocates and NGOs. A harsh condemnation of Italy was also delivered by Dunja Mijatović, Commissioner for Human Rights at Council of Europe[12]. From an epidemiological perspective, this measure appeared to be nonsense because it confined and combined in a limited space people who had no symptoms of infection with other people who were found to be COVID-19 positive. This endangered the health of the former group and prevented the second group from receiving adequate care, which could have prevented virus transmission on the ferries. Such a confinement – which resembles a deportation – not only contradicted the basic principles of medical ethics and any sense of humanity; it led to the harm of vulnerable people such as migrants and asylum-seekers,

---

9 Translated from the original Italian.

10 *Bando per 4 navi-quarantena da 300–400 posti*, May 17, 2021, available from: https://www.ansa.it/sito/notizie/topnews/2021/05/17/migranti-bando-per-4-navi-quarantena-da-300-400-posti_35d2259d-f2b9-4fc5-83d5-ecd91514bd41.html.

11 This was declared at the beginning of pandemic to offer the government extra power to manage the outbreak of the virus.

12 See The New Humanitarian. Italy's use of ferries to quarantine migrants has come under fire https://www.thenewhumanitarian.org/news-feature/2020/11/9/italy-migration-ferries-coronavirus-quarantine-health-asylum.

who, as seen previously, ordinarily experience trauma and suffering throughout their migration routes. Paradoxically, the costs of this measure have been reported to be much higher than the costs that would have been required to provide full assistance to migrants on land[13].

The use of commercial ferries to manage illegal migrations flows during the health emergency must be linked to the attempts to criminalize the search and rescue (SAR) operations conducted by NGO boats in the Mediterranean Sea area, particularly on the route from Libya. The activity of NGOs in SAR operations progressively increased in 2014 due to the end of "Mare Nostrum" a program funded by the Italian Government to save the lives of migrants rescued in the Mediterranean Sea (Masera 2018). After the program's termination, NGOs activities unofficially came under the management of the SAR activities of the Italian naval authorities, as shown by the fact that from July to December 2016, approximately 40% of the SAR operations were conducted by NGOs boats coordinated by the Italian Coast Guard (Masera 2018).

The situation abruptly changed in 2017, after the publication of a report by Frontex – the EU force responsible for border control – according to which SAR operations conducted by NGOs were described as a "pull factor" for migrations flows. In other words, NGOs were depicted as a "facilitator" of illegal migration routes and therefore of migrant smuggling. Soon thereafter, a public prosecutor from Catanzaro evoked the connections between smugglers and NGOs (Masera 2018), thus triggering Italian right-wing political parties to foster indignation among the public opinion. Due to this situation, the Italian Government decided to hamper the efforts of NGOs by asking them to sign a "Conduct Code for NGOs committed to rescuing migrants in the sea"[14] as a condition to conducting SAR operations in the Mediterranean Sea. Many NGOs protested this imposition and refused to sign the Conduct Code, particularly because it required them to accept the presence of police officers on board their boats and to cooperate with the Libyan Coast Guard (Arpinati 2021). As a result, several NGO boats were prevented entry to Italian ports and sequestrated by the Italian authorities, and members of the NGOs were prosecuted for facilitating illegal migration. However, most of these charges were dropped during the investigation stage (Arpinati 2021).

---

13 AGI Press, *Quanto costano allo Stato le navi quarantena per i migranti*, July 15, 2020. Available: https://www.agi.it/cronaca/news/2020-07-15/costo-stato-navi-quarantena-migranti-9164639/.
14 The Conduct Code was issued by the former Ministry of Interior in August 2017. Available: https://www.interno.gov.it/sites/default/files/codice_condotta_ong.pdf.

5 Between exploitation and criminalization

The attempts by the Italian Government to criminalize SAR operations conducted by the NGOs in the Mediterranean Sea must be framed in the context of the wider EU strategy for "managing flows" (Botrugno 2014b). Among other goals, the latter is intended to increase the capacity of EU neighbor countries to "filter" migration flows and hold back a large number of migrants. The bilateral agreements signed by the EU with the governments of third countries under the auspices of the European Neighbourhood Policy have made way for an "externalization" of the borders. Namely, the national authorities of third countries have been urged by the EU to prevent illegal migrants' departure with the aim of reducing incoming flows (Botrugno 2014b). However, many of the countries with which the EU and the Member States have signed partnerships completely disregard the protection of fundamental rights. This explains why the local authorities of these countries are often charged of inflicting on migrants in transit abuses and violence that do not differ from those inflicted on them by traffickers and smugglers (FTDES 2021; Mediterranea 2022; Border Violence Monitoring Network 2022).

When intercepted at the EU's land borders or rescued in the Mediterranean Sea, migrants are brought to "identification centers" – which must be read as detention centers – where they are forced to stay for several months (ranging from 6 to 18 months) in conditions that are comparable to those of the Nazi lagers. On paper, their residence in these centers should not extend beyond the time required for their reception, identification, and eventual repatriation. However, this physical containment has progressively become a pillar of EU flow management strategies.

The European Committee for the Prevention of Torture and Inhuman or Degrading Treatment or Punishment (CPT) has repeatedly emphasized that most of the detention centers throughout Europe are overcrowded and are deficient with respect to the most basic hygienic conditions (CPT 2017). Moreover, as has been reported by several NGOs (MSF 2016; MEDU 2013), this form of detention is pathogenic and human rights violations have been repeatedly reported. Moreover, the centers often host rival ethnic groups in the same space, increasing the likelihood of fights and violence among detainees (MEDU 2013). Consequently, diseases, violence, and traumas easily proliferate among the detainees, leading to a worrying escalation of episodes of self-harm and suicide (Fekete 2011; MSF 2016; MEDU 2017). Most of these centers also lack adequate care professionals and services, and drug abuse has been reported as a means of controlling detainees and preventing agitations. For instance, a study conducted by the Italian Association MEDU in 2013 showed that prolonged stays in detention centers are often associated with initiating or increasing use of benzodiazepines (MEDU 2013). These drugs are often administered without due psychiatric assessment and at higher dosages than recommended – up to seven times higher – thus leaving people in a state of confusion for a prolonged time. Drug use has also been reported as a way of punishing detainees that try to escape, exhibit dissent, or adopt attitudes that are considered to be inappropriate by personnel working at the centers (MEDU 2013).

The Italian National Bioethics Committee (Comitato Nazionale per la Bioetica 2017) confirmed that the conditions of these centers are inhumane and incompatible with the need to protect the lives and health of migrants and asylum-seekers; thus, the Committee recommended their closure. Moreover, the Committee asked the introduction of the crime of torture in Italy with the aim of preventing "dramatic experiences like those suffered by migrants, especially women, such as arbitrary detention, inhuman treatments, sexual harassment and rape, and slavery for purpose of prostitution"[15] (Comitato Nazionale per la Bioetica 2017, p. 4).

Migrants who escape from the control of border authorities are forced to live in a precarious and uncertain condition, which often leads them to resort to the black labor market or to fall into the trap of criminality. As described by Lucia Re in the first chapter of this book (see Section 2), the presence of foreign citizens in EU penitentiaries is disproportionate when correlated with their share among the general population. This confirms that the difficulties experienced by migrants with respect to obtaining a legal status and regular employment function as the waiting room of criminalization processes. In this regard, Emilio Santoro already noted that the deliberate exclusion of migrants from citizenship and the enjoyment of social rights is used to prevent the "perception of the inevitable scarcity of resources at the disposal of the state for welfare purposes"[16] (Santoro 2006, p. 69). From this perspective, although migrants are largely denied civil and social rights, they are massively exploited by the productive systems of the EU countries and used at the political level as a scapegoat to justify the progressive loss of control over employment dynamics by national governments.

The management of the pandemic has confirmed the trend toward the exploitation of migrants' precarious and vulnerable conditions (Sanfelici 2021). The only attempt to enhance the protection of undocumented migrants during the pandemic has been the approval of a "regularization" reform in 2020 (Italian

---

15 Translated from the original Italian.
16 Translated from the original Italian.

Government Decree no. 34/2020). This reform offered employers the opportunity to regularize illegally hired Italian or foreign workers in the primary sector (i.e. agriculture, farming, fishing, zootechnics) or in domestic and care work. This process was meant to allow undocumented migrants to obtain a legal status, albeit a temporary one. However, despite the emphasis of the Italian Government in the context of launching the reform[17], a very limited number of migrants were able to access this benefit due to its severe requirements. On this account, the regularization program encompassed several critics (Zorzella 2020; Santoro 2021; Caprioglio and Rigo 2020) as it was deemed to be insufficient to provide protection to a large number of undocumented migrants who, although they are criminalized and marginalized, represent a fundamental workforce for the Italian economy (Sanfelici 2021). Many of the undocumented migrants employed in the black labor market, particularly those working in the agriculture or construction sectors, are currently forced to live in "ghettos", i.e. run-down shelters located in abandoned peripheral sites, most of which lack access to electricity and clean water. Regarding healthcare access, these individuals do not receive assistance other than that provided by local NGOs. Italian scholars have described the conditions of these people as a form of "modern slavery" (Torre 2019; Santoro 2009) given they are literally worked to death, as happens much too often.

6 A stranger in a strange land: the burden of socio-cultural misconceptions in health and healthcare

In "The Foundation of Bioethics", Tristam H. Engelhardt coined the now-famous expression "a stranger in a strange land" to evoke the sense of alienation experienced by patients disentangling the complex organization of contemporary healthcare systems (Engelhardt 1996, p. 295). Although the bioethicist was not referring to migrants, this expression can be used metaphorically to understand the feelings of hostility suffered by many migrants – undocumented migrants in particular – when accessing or seeking to access healthcare services in destination countries. Beyond their lack of a legal status, which, as shown above, can be a main obstacle, other factors can prevent migrant people from receiving ade-

---

17 Announced by the former Minister of Agriculture, Teresa Bellanova, as a historical step for overcoming labour exploitation in Italy.

quate care. Language barrier can be a severe obstacle for migrants, particularly if point-to-care services are not prepared to offer multilingual assistance. Mutual incomprehension and misunderstandings can amplify the negative feelings of strangeness and unfamiliarity experienced by migrants in the context of the healthcare system. In this regard, it is worth remembering that, despite the efforts made in the fields of bioethics and medical ethics to foster a "therapeutic alliance", there remains a distance between the "technical perspective" of healthcare professionals and the "lay vision" of patients. Clark and Mishler (1992) have described this distance as a "conflict" between two different "voices": "the 'voice of medicine', expressing a technical, biomedical frame of reference, and the 'voice of the lifeworld', reflecting the patient's personal, 'contextually-grounded experiences of events and problems', expressed in familiar terms" (Clark and Mishler 1992, p. 346). As they further argue, "[u]sually, the voice of medicine dominates the discourse, but the conflict tends to recur throughout the encounter at various levels of intensity" (Clark and Mishler 1992, p. 346).

From this perspective, the efforts made by patients to tell their own "stories" must not be considered mere complaints that accompany the suffering experienced; rather, they represent attempts to "make sense" of technical elements such as symptoms and medical prescriptions. If healthcare professionals are unaware of this distance and thus do not strive to find a balance between the two "voices", patients – not only migrant patients – are likely to have a negative experience with healthcare delivery. This can even undermine the efficacy of healthcare and induce negative feelings such as fear, discomfort, and inadequacy in people who are vulnerable due to their health conditions.

Another potential obstacle related to communication between patients and healthcare professionals is culture. Behind words are social and cultural practices, personal and collective values, beliefs, and especially norms and customs. Due to the asymmetrical relationship between patients and healthcare professionals – and the feelings of "strangeness" experienced by many migrants in such a context – a fundamental part of the relationship could be absorbed by nonverbal communication. The same holds true of verbal communication; even the latter is a socially and culturally grounded mode of expression of feelings and thoughts. The ways in which sensations such as pain, fear, embarrassment, approval, disagreement, (dis)satisfaction, sadness, happiness, subjection, or reverence are expressed or hidden are strongly dependent on the cultural variables that shape individual and collective identities.

These multiple forms must be coupled with similarly multiple and culturally-informed ways of conceptualizing the body as well as its relations with the "world", i.e. the environment, the society and its structures. Despite the fact that, on paper, bioethics is well-equipped to deal with conflicts and find a balance among competing interests, cultural fragmentation of industrialized societies demands bioethics scholars to be "culturally competent", as they are increasingly required to determine, on a case-to-case basis, what belongs to the moral spectrum of individuals and social groups, what the individual's hierarchy of values is, and how to balance dominant hierarchies with those of minorities (Mocellin 2011; see also Lorenzo and Mocellin Raymundo in this book). Of course, all this effort must feature the collaboration of patients themselves. Insofar as healthcare professionals and the healthcare systems in which they operate overlook the relevance of this work – i.e. because they believe that medicine is unrelated to culture – not only does healthcare become difficult to access for the migrant population, but the risk is that cultural differences could be ignored or "pathologized" as a deviation from standards (Botrugno 2014c).

7 Structural racism and discrimination: From inequalities to invisibility

As mentioned above (Section 1), even before the COVID-19 outbreak, a considerable body of evidence showed that the status of "migrant" represents a significant disadvantage with respect to both maintaining a good health condition and gaining access to adequate healthcare in destination countries. The countless inequalities reported on the part of migrants include but are not limited to a higher rate of complications in pregnancy and birth, and lower access to gynecological public services than native women (Tognetti Bordogna 2013); more difficulties expressing individual needs and understanding physicians than native patients (European Commission 2013); higher exposure to on-the-job injuries and job-related diseases than the native population – up to two times higher – while migrants' levels of employment are lower (EU-OSHA 2007; Reyneri 2011); higher incidence of post-traumatic stress disorders among refugees – approximately ten times higher – than among the general population (Fazel et al. 2005); and higher rates of depression and anxiety disorders among refugees – approximately two times higher – than among general population (Lindert et al. 2009). In 2011, the European Parliament adopted a statement on "Reducing Health Inequalities" (EU Parliament 2011), in which it called member states

> to ensure that the most vulnerable groups, including undocumented migrants, are entitled to and are provided with equitable access to healthcare [and] to assess the feasibility of supporting healthcare for irregular migrants by providing a definition based on common

principles for basic elements of healthcare as defined in their national legislation (EU Parliament 2011, p. 5).

Specific consideration was given to the protection of the health of migrant women and the resulting necessity of developing "training initiatives enabling doctors and other professionals to adopt an intercultural approach based on recognition of, and respect for, diversity and the sensitivities of people from different geographical regions" (EU Parliament 2011, p. 6).

The COVID-19 pandemic has exacerbated inequalities in health and healthcare and widened the gaps between the native and foreign populations. To a large extent, the perpetration of such inequalities seems to have been driven by the adoption of discriminatory policies that have excluded migrants from access to healthcare services or made such access more difficult (Mangesha et al. 2022; Crawshaw et al. 2022; Machado and Goldenberg 2021; De Berardi et al. 2021; Mukumbang 2021; Worthing et al. 2021). In this regard, Machado and Goldenberg emphasized that "[g]lobal pandemic responses have neglected im/migrants by continuing to ignore or at best, insufficiently address inequities, exacerbating COVID transmission, xenophobia, racism, and occupational injustice" (Machado and Goldenberg 2021, p. 2). From this perspective, "[d]eaths, illness, stress, and other negative consequences of overlapping issues of COVID-19 and precarious im/migration status highlight the tangible, life-threatening manifestation of these inequities, perpetuating structural racism" (Machado and Goldenberg 2021, p. 2).

In some cases, the exacerbation of these inequalities was not an indirect effect but rather a form of deliberate discrimination, as in cases featuring no or only a delayed inclusion of undocumented migrants and asylum-seekers in the COVID-19 vaccination program (De Berardi et al. 2021, p. 2; Crawshaw et al. 2022, p. 9)[18]. As seen above (Section 4), the exclusion of vulnerable populations such as displaced people from vaccination can foster transmission of the virus, putting the health of the whole society at risk. In addition, migrants are particularly exposed to the risk of vaccine hesitancy, not only with respect to COVID-19 but also in the case of routine vaccines. A systematic review conducted by Crawshaw and colleagues confirmed the risk of vaccine hesitancy among migrants, particularly among groups that have been disproportionately affected by the virus: "hesitancy in these populations might be an expression of cultural ali-

---

18 Among these countries is Italy, where undocumented migrants were first included in the vaccination program only in September 2021. See https://www.ilsole24ore.com/art/decreto-green-pass-bis-migranti-irregolari-vaccinati-certificato-provvisorio-AEaaeyk.

enation resulting from experiences of marginalisation or discrimination" (Crawshaw et al. 2022, p. 9). In this context, the review shed light on the existence of "barriers from gaps in health-care provider knowledge around catch-up vaccination, an area where experts have called for more guidelines" (Crawshaw et al. 2022, p. 9).

As seen previously, besides being discriminatory and totally contradicting the most basic ethical principle of caring for the sick, these politics are epidemiologically unfounded and economically disproportionate. In addition, they could be dangerous even for native populations given that they cause migrants to hide their health conditions in an attempt to elude border controls. As suggested by a substantial body of literature in this field (e. g. Soto 2009; IOM 2010; WHO 2013), the key to ensuring safety against the spread of infectious diseases is the implementation of adequate surveillance systems and specific screening programs aimed at detecting and neutralizing major threats to public health. In this regard, Crawshaw and colleagues (2022) argue that healthcare systems should improve their capacity to screen, which must be read as "seeking migrants in need of care" instead of closing their eyes to their situation and suffering. To achieve this goal, it is necessary to "tackle the systemic barriers to accessing vaccination by creating more culturally competent health systems" (Crawshaw et al. 2022, p. 9). Migrants have indeed reported a lack of trust towards the healthcare system and shown multiple difficulties communicating with healthcare professionals and accessing or understanding vaccination information. This caused them "to avoid care, delay vaccination, or turn to alternative sources, including social media" (Crawshaw et al. 2022, p. 9). In turn, healthcare professionals have emphasized "the additional burden that communication barriers and lack of interpreters imposed on their limited consultation time" (Crawshaw et al. 2022, p. 9).

However, hitherto, only a few EU member states offer immunization programs to benefit migrants and refugees (WHO 2017b), which means that the latter groups continue to face significant obstacles with respect to accessing vaccination services in destination countries.

Accordingly, it is possible to argue that, in industrialized societies, migrants are often caught in a "pathological circle" made of discrimination, criminalization, language barriers, poor living conditions, and socio-cultural misconceptions (McGuire & Martin 2007; Fassin 2001). In this circle, migrants are often made "invisible" given that they are discriminated against to a large extent and thus have a reduced level of agency compared to native population, including a reduced ability to navigate the bureaucratic organizations of healthcare systems. In addition, they have limited time and limited economic resources, which are linked with the fear of being discriminated against and blamed when seeking access to healthcare services. This affects their agency and, once again, condemns them to invisibility in destination societies.

8 Conclusions

According to Marcia M. Raymundo (2011), bioethics must be able to recognize and protect a plurality of epistemologies and cultural perspectives in healthcare, avoiding the dominance of any particular viewpoint – usually the western biomedical perspective – over the others. In practical terms, this would entail developing a set of tools to help individuals reacting to the hierarchies of values, interests, and practices resulting from established relations of power in the context of healthcare. Such an orientation is compliant with the 2001 UNESCO "Declaration on Cultural Diversity", according to which defending cultural diversity should be considered "an ethical imperative, inseparable from respect for human dignity" (UNESCO 2001, art. 4). Accordingly, defending this imperative implies "a commitment to human rights and fundamental freedoms, in particular the rights of persons belonging to minorities and those of indigenous peoples. No one may invoke cultural diversity to infringe upon human rights guaranteed by international law, nor to limit their scope" (UNESCO 2001, art. 4).

This perspective of bioethics fits the idea of "everyday bioethics" proposed by Giovanni Berlinguer (Berlinguer 2000; 2003), which was not meant to be a new or alternative theory of bioethics. Rather, it was described as a perspective aimed at revitalizing the importance of elements such as discrimination, racism, and (in)equity, which have always belonged to the domain of bioethics, but have been relegated to the background due to the prevalence of more individual-based theories, approaches, and research topics. In other words, enhancing an everyday bioethics can be viewed as an attempt to foster social justice in bioethics. When I reported the proliferation of "zero tolerance" policies aimed at preventing access to healthcare by undocumented migrants (Martinez et al. 2015; see also Section 2) above, I also emphasized the fact that many healthcare professionals, NGOs, and other people have responded to those inhumane and unfair rules, which they felt were not grounded in common morality. Countless healthcare professionals, in particular, have deliberately disregarded those provisions and defended everybody's right to access healthcare, regardless of their status. The ethical duty to care for the sick stands at the foundation of modern medicine itself and is also a core of the contemporary version of the Hippocratic Oath, which serves as a bridge between healthcare practice and (bio)ethics. These professionals' disobedience contributed to opening the black box of law, showing the material gaps, the conflicts, and the negotiations that separate the "law in the books" from the "law in action". From this perspective, everyday bioethics can also be regarded as a form of "bioethics in action".

Enhancing everyday bioethics also requires further research and further education based on this perspective. In the context of migrants, this entails educating healthcare professionals on ways of receiving foreign patients and establishing good (intercultural) care relationships. It also involves raising awareness among the native population of the fact that migration flows are not a "disease" of the contemporary era, nor does the migrants' arrival represent a crisis or a threat to our stability. Rather, migration flows are a combination of poverty, war, persecution, and social exclusion, something for which industrialized countries have enormous responsibilities (Buxo i Rey 2004). Native populations should be aware that migrants are usually healthier than people who are born in industrialized countries (e.g. Constant et al. 2017; Kennedy et al. 2015; Domnich et al. 2012) and that their health problems are mostly due to malnutrition, poor education, and poor living conditions. In addition, migrants represent a significant resource for the economic development of destination countries (Botrugno 2014b), as has been widely acknowledged by EU policy-makers (European Commission 2005).

To summarize, there are several reasons why we cannot hesitate to care for migrants' health. First, it's a matter of public health: effective prevention requires the implementation of surveillance systems and immunization programs to neutralize the potential threat of spreading diseases. Second, it's a matter of professional deontology: healthcare professionals have an ethical obligation to care for the sick, which implies not "closing one's eyes" in the face of human suffering but rather developing a proactive attitude to meet the needs of the most vulnerable and underserved populations. Third, and most important, it's a matter of justice: migration flows should be viewed as a global movement for social justice, and healthcare access is a fundamental right that must be protected regardless of citizenship or legal status.

To ensure that migrants have full access to healthcare, it is necessary to adopt policies that are aimed at compensating for health inequalities and eliminating discrimination in healthcare, which may be achieved through the promotion of multilingual migrant-friendly care services, the development of a systemic approach to migrants' health protection that should be shared among all healthcare systems in the EU, and the education of healthcare professionals to enhance the recognition and protection of cultural diversity and to combat "policies of deterrence" and anti-immigrant rhetorics.

In conclusion, a "bioethics in action" offers a substantial contribution to the task of raising public opinion and healthcare professionals' awareness concerning the importance of combatting inequalities, striving for equity, and protecting the fundamental rights of all populations, especially migrant populations, which are some of the most vulnerable.

References

ASGI. (2021): "L'esperimento delle navi quarantena e i principali profili di criticità". https://www.asgi.it/allontamento-espulsione/navi-quarantena-esperimento/, visited on 20 June 2022.

Arnold, Franco; Katona, Cornelisu; Cohen, Juliet; Jones, Lucy; and McCoy, David. (2015): "Responding to the needs of refugees". In: *BMJ* 351, h6731. doi: 10.1136/bmj.h6731.

Beaunoyer, Elisabeth; Dupéré, Sophie; and Guitton, Matthieu J. (2020): "COVID-19 and digital inequalities: Reciprocal impacts mitigation strategies". In: *Computers in Human Behavior* 111. N. 106424, doi: 10.1016/j.chb.2020.106424.

Berardi, Chiara; Lee, Eun Su; Wechtler, Heidi; Paolucci, Francesco. (2021): "A vicious cycle of health (in)equity: Migrant inclusion in light of COVID-19". In: *Health Policy Technol* 11. N. 2, 100606, doi: 10.1016/j.hlpt.2022.100606.

Berlinguer, Giovanni. (2003): *Everyday Bioethics: Reflections on Bioethical Choices.* New York: Baywood Publisher.

Berlinguer, Giovanni. (2000): *Bioetica quotidiana.* Firenze: Giunti.

Bhugra, Dinesh; and Gupta, Susham [Eds]. (2010): *Migration and Mental Health.* Cambridge: University Press.

Border Violence Monitoring Network. (2022): *Illegal push-backs and border violence report.* https://www.borderviolence.eu/wp-content/uploads/BVMN-Monthly-Report-January-2022.pdf-2.pdf, visited on 20 June 2022.

Botrugno, Carlo. (2014a): "Il diritto alla salute dinnanzi alla sfida della complessità: dalla crisi del riduzionismo biomedico alla global health". In: *Italian Journal of Legal Philosophy* 2. Pp. 495–512.

Botrugno, Carlo. (2014b): "Immigrazione e Unione Europea: un excursus storico ragionato". In: *Sociologia del diritto* 1. Pp. 143–65.

Botrugno, Carlo. (2019a): "Uno sguardo all'intersezione: migrazione, disabilità e lotta per i diritti". In Bernardini, Maria Giulia (Ed), *Migrants with Disability and Vulnerability. Representations, Policies and Rights*, pp. 95–116. Jovene: Naples.

Botrugno, Carlo. (2019b): "Diritto alla salute e migrazioni internazionali: per una bioetica in azione". In: *Jura Gentium* 16. N. 2, pp. 102–26.

Botrugno, Carlo. (2020a): "El papel de la tecnología en la gestión de la pandemia de CoViD-19". In: *RedBioetica Unesco* 21, pp. 13–20.

Botrugno, Carlo. (2020b): "CoViD-19, nuove tecnologie e diritti fondamentali". In: *L'Altro Diritto* 4. Pp. 21–39.

Botrugno, Carlo; Kaplan, Bonnie; Di Bartolomeo, Gabrielle. (2022): "Ethical, legal, and social issues in digital dermatology". In: Nouri, Keyan Ed., *Telemedicine and technological advancements in dermatology.* Springer Nature (in press).

Botrugno, Carlo. (2018): "Healthcare, migrations and everyday bioethics: Weighing the difference". In: *L'altro Diritto* 1. Pp: 91–118.

Bourdieu, Pierre. (1977): *Outline of a Theory of Practice.* Cambridge: Cambridge University Press.

Bourdieu, Pierre. (1986): "The forms of capital". In: Richardson, John G. (ed), *Handbook of Theory and Research for the Sociology of Education*, pp. 241–258. New York: Greenwood.

Buffa, Matteo; Casadei, Thomas; Lettieri, Nicola; Magneschi, Chiara, and Sciurba, Alessandra. (2021): "Appello. Chiarezza sulle 'navi quarantena'. No alla contrapposizione tra tutela della salute pubblica e rispetto dei diritti fondamentali". In: *L'altro Diritto* 1. Pp: 135–137.

Buxó i Rey, Maria Jesús. (2004): "Bioética Intercultural para la Salud Global". In: *Revista de Bioética y Derecho* 1. Pp: 12–5.

Caprioglio, Carlo; and Rigo, Enrica. (2020): "Lavoro, politiche migratorie e sfruttamento: la condizione dei braccianti migranti in agricoltura". In: *Diritto, immigrazione e cittadinanza* 3. Pp: 33–56.

Castelli, Franco; and Sulis, Giovanni. (2017): "Migration and infectious diseases". In: *Clinical Microbiology and Infection* 23. N. 5, pp. 283–289.

Clark, Jack A.; and Mishler, Elliot G. (1992): "Attending to patients' stories: referencing the clinical task". In: *Sociology of Health and Illness* 14. Pp. 344–372.

Comitato Nazionale per la Bioetica. (2010): *Le condizioni di vita della donna nella terza e quarta età: aspetti bioetici nella assistenza socio-sanitaria.* http://bioetica.governo.it/media/170708/p93_2010_donne_terza_quarta_eta_it.pdf, visited on 23 July 2022.

Comitato Nazionale per la Bioetica. (2017): *Immigrazione e salute.* http://bioetica.governo.it/it/pareri/pareri-e-risposte/immigrazione-e-salute/, visited on 23 July 2022.

Constant, Amelie F.; García-Muñoz, Teresa; Neuman, Shoshana; and Neuman, Tzahi. (2017): "A "Healthy Immigrant Effect" or a "Sick Immigrant Effect"? Selection and Policies Matter". In: *Eur J Health Econ* 19. N. 1, pp. 103–121.

CPT [European Committee for the Prevention of Torture and Inhuman or Degrading Treatment or Punishment, Council of Europe]. (2017): *Immigration Detention Factsheet.* https://rm.coe.int/16806fbf12, visited on 23 July 2022.

CPT [Committee for the Prevention of Torture and Inhuman or Degrading Treatment or Punishment, Council of Europe]. (2020): *Statement of principles relating to the treatment of persons deprived of their liberty in the context of the coronavirus disease (COVID-19) pandemic.* https://rm.coe.int/16809cfa4b, visited on 23 July 2022.

Crawshaw, Alison F.; Farah, Yasmine; Deal, Anna; Rustage, Kieran; Hayward, Sally E.; Carter, Jessica; Knights, Felicity; Goldsmith, Lucy P.; Campos-Matos, Ines; Wurie, Fatima; Majeed, Azeem; Bedford, Helen; Forster, Alice S.; Hargreaves, Sally. (2022): "Defining the determinants of vaccine uptake and undervaccination in migrant populations in Europe to improve routine and COVID-19 vaccine uptake: a systematic review". In: *Lancet Infect Dis* S1473–3099(22)00066–4.

Crouzet, Lisa; Scarlett, Honor; Colleville, Anne-Claire; Pourtau, Lionel; Melchior, Maria; Ducarroz, Simon. (2022): "Impact of the COVID-19 pandemic on vulnerable groups, including homeless persons and migrants, in France: A qualitative study". In: *Prev Med Rep* 26. N. 101727, doi: 10.1016/j.pmedr.2022.101727.

Cuttitta, Paolo. (2007): "Le rotte che cambiano i confini". In: *Segno* 33. N. 289, pp. 49–54.

Domnich, Alexander; Panatto, Donatella; Gasparini, Roberto; Amicizia, Daniela. (2012): "The 'healthy immigrant' effect: does it exist in Europe today?". In: *The Italian Journal of Public Health* 9. N. 3, pp. 1–7.

Engelhardt, Hugo Tristam. (1996): *The Foundations of Bioethics*. New York: Oxford University Press.

EPHA [European Public Health Alliance]. (2014): *Health inequalities and eHealth. Report of the eHealth Stakeholder Group*. https://ec.europa.eu/digital-single-market/news/commis sion-publishes-four-reports-ehealth-stakeholder-group, visited on 13 July 2022.

EU-OSHA [European Agency for Safety and Health at Work]. (2007). *Annual report 2007: bringing safety and health closer to European workers*. https://osha.europa.eu/en/ publications/annual_report/2007full/view, visited on 13 July 2022.

European Commission. (2005): *The Green Paper on an EU approach to Managing Economic Migration*. europa.eu/rapid/press-release_SPEECH-05-364_en.pdf, visited on 13 July 2022.

European Commission. (2013): *Health inequalities in the EU. Final Report of a Consortium*. https://ec.europa.eu/health//sites/health/files/social_determinants/docs/health inequalitiesineu_2013_en.pdf, visited on 13 July 2022.

European Union Parliament. (2011): *Reducing health inequalities in the EU*. Resolution 8 March 2011, 2010/2089(INI). http://www.europarl.europa.eu/sides/getDoc.do?pubRef=-// EP//NONSGML+TA+P7-TA-2011-0081+0+DOC+PDF+V0//EN, visited on 13 July 2022.

Farmer, Paul. (2004): "An anthropology of structural violence". In: *Current Anthropology* 45. N. 3, pp. 305–325.

Fassin, Didier. (2001): "Une double peine. La condition sociale des immigrés malades du sida". In: *L'Homme* 160. Pp. 137–162.

Fazel, Mina; Wheeler, Jeremy; Danesh, John. (2005): "Prevalence of serious mental disorder in 7000 refugees resettled in western countries: a systematic review". In: *The Lancet*, 9–15. N. 365(9467), pp. 1309–1314.

Fekete, Liz. (2011): "Accelerated removals: the human cost of EU deportation policies". In: *Race and Class* 52. N. 4, pp. 89–97.

Feng, Yang; Xie, Wenjing. (2015): "Digital divide 2.0: the role of social networking sites in seeking health information online from a longitudinal perspective". In: *J Health Commun* 20. N. 1, pp. 60–68. doi: 10.1080/10810730.2014.906522.

FTDES. (2021): *La Tunisie, porte de l'Afrique & frontière de l'Europe Rapport d'une mission de recherche entre Sfax, Zarzis et Medenine*. https://ftdes.net/rapports/frontiereseurope. pdf, visited on 13 July 2022.

Fundamental Rights Agency. (2011): *Migrants in an irregular situation employed in domestic work: Fundamental rights challenges for the European Union and its Member States*. http://fra.europa.eu/en/publication/2012/migrants-irregular-situation-employed-domes tic-work-fundamental-rights-challenges, visited on 13 July 2022.

Fundamental Rights Agency. (2017): *European Union minorities and discrimination survey*. http://fra.europa.eu/en/project/2015/eu-midis-ii-european-union-minorities-and-discrim ination-survey, visited on 13 July 2022.

Gama, Ana; Rocha, João Victor; Marques, Maria J.; Azeredo-Lopes, Sofia; Pedro, Ana Rita; Dias, Sónia. (2022): "How Did the COVID-19 Pandemic Affect Migrant Populations in Lisbon, Portugal? A Study on Perceived Effects on Health and Economic Condition". In: *Int J Environ Res Public Health* 19. N. 3, 1786. doi: 10.3390/ijerph19031786.

Garattini, L., Zanetti, M. and Freemantle, N. (2020): "The Italian NHS: What Lessons to Draw from CoViD-19?". In: *Appl Health Econ Health Policy* 18. N. 4, pp. 463–466. doi: 10.1007/s40258-020-00594-5.

Gianguzza, Giulia; El Karkouri, Kamal. (2020): "El babour: il modello delle 'navi quarantena' e il suo impatto sulla vita delle persone trattenute a bordo". In: *L'altro Diritto* 1. Pp. 120–134.

Greenaway, Christina; Hargreaves, Sally; Barkati, Sapha; Coyle, Christina M., Gobbi, Federico; Veizis, Apostolos; Douglas, Paul. (2020): "COVID-19: Exposing and addressing health disparities among ethnic minorities and migrants". In: *J Travel Med* 27. N. 7, taaa113. doi: 10.1093/jtm/taaa113.

Gruer, Laurence; Agyemang, Charles; Bhopal, Ray; Chiarenza, Antonio; Krasnik, Allan; Kumar, Bernadette (2021): "Migration, ethnicity, racism and the COVID-19 pandemic: A conference marking the launch of a new Global Society". In: *Public Health Pract* 2. N. 100088, doi: 10.1016/j.puhip.2021.100088.

Hollander, Judd E.; Brendan, Carr G. (2020): "Virtually Perfect? Telemedicine for Covid-19". In: *N Engl J Med* 382. Pp: 1679–1681.

IOM [International Organization for Migration]. (2017): *Missing Migrants Project Report.* https://missingmigrants.iom.int/region/mediterranean, visited on 13 July 2022.

IOM [International Organization for Migration]. (2010): *Increasing Public Health Safety Alongside The New Eastern European Border: An Overview of Findings from the Situational Analysis.* http://www.iom.int/jahia/webdav/shared/shared/mainsite/activities/health/PHBLM-SAR-Public-Report-2010.pdf, visited on 13 July 2022.

IOM [International Organization for Migration]. (2020): *World migration report.* https://world migrationreport.iom.int/wmr-2022-interactive/, visited on 13 July 2022.

ISS [Istituto Superiore Sanità]. (2015) Salute pubblica ed emergenza immigrazione, available at: http://www.epicentro.iss.it/argomenti/migranti/ReportSpeim.asp.

Jaljaa, Anissa; Caminada, Susanna; Tosti, Maria Elena; D'Angelo, Franca; Angelozzi Aurora; Isonne, Claudia; Marchetti, Giulia; Mazzalai, Elena; Giannini, Daria; Turatto, Federica; De Marchi, Chiara; Gatta, Angela; Declich, Silvia; Pizzarelli, Silla; Geraci, Salvatore; Baglio, Giovanni; Marceca, Marceca. (2022): "Risk of SARS-CoV-2 infection in migrants and ethnic minorities compared with the general population in the European WHO region during the first year of the pandemic: a systematic review". In: *BMC Public Health* 22. N. 1, 143. doi: 10.1186/s12889-021-12466-1.

Kaplan, Bonnie. (2022): "Ethics, guidelines, standards, and policy: Telemedicine, COVID-19, and broadening the ethical scope". In: *Camb Q Healthc Ethics* 31. N. 1, https://doi.org/10.1017/S0963180121000852.

Kaplan, Bonnie. (2020): "Revisiting Health Information Technology Ethical, Legal, and Social Issues and Evaluation". In: *JMIR*, ahead of print, https://doi.org/10.1016/j.ijmedinf.2020.104239.

Kawachi, Ichiro. (1999): "Social capital and community effects on population and individual health". In: *Annals of the New York Academy of Sciences* 896. Pp. 120–30.

Kennedy, Steven; Kidd, Michael P.; McDonald, James Ted; Biddle, Nicolas. (2015): "The Healthy Immigrant Effect: Patterns and Evidence from Four Countries". In: *Journal of International Migration and Integration* 16. 2, pp. 317–32.

Knights, Felicity; Carter, Jessica; Deal, Anna; Crawshaw, Alison F.; Hayward, Sally E.; Jones, Lucina; Hargreaves, Sally. (2021): "Impact of COVID-19 on migrants' access to primary care and implications for vaccine roll-out: a national qualitative study". In: *Br J Gen Pract* 71. N. 709, pp. e583-e595. doi: 10.3399/BJGP.2021.0028.

Kontos, Emily; Blake, Kelly D.; Wen-Ying, Sylvia C.; Prestin, Abby. (2014): "Predictors of eHealth usage: insights on the digital divide from the Health Information National Trends Survey 2012". In: *JMIR* 16. N. 7, e172. doi: 10.2196/jmir.3117.

Latulippe, Karine; Hamel, Christine; Giroux, Dominique. (2017): "Social Health Inequalities and eHealth: A Literature Review With Qualitative Synthesis of Theoretical and Empirical Studies". In: *JMIR* 19. N. 4, e136. doi: 10.2196/jmir.6731.

Lebret, Audrey. (2020): "COVID-19 pandemic and derogation to human rights". In: *Journal of Law and the Biosciences* 7. N. 1, pp. 1–15.

Lindert, Jutta; von Ehrenstein, Ondine S.; Priebe, Stefan; Mielck, Andreas; Brähler, Elmar. (2009): "Depression and anxiety in labor migrants and refugees – A systematic review and meta-analysis". In: *Social Science and Medicine* 69. Pp. 246–57.

Machado, Stefanie; Goldenberg, Shira. (2021): "Sharpening our public health lens: advancing im/migrant health equity during COVID-19 and beyond". In: *Int J Equity Health* 20. N. 1, 57. doi: 10.1186/s12939-021-01399-1.

Mangrio, Elisabeth; Zdravkovic, Slobodan; Strange, Michael. (2022): "Working With Refugees' Health During COVID-19-The Experience of Health- and Social Care Workers in Sweden". In: *Front Public Health* 10. N. 811974. doi: 10.3389/fpubh.2022.811974.

Markel, Howard; Stern, Alexandra M. (2002): "The Foreignness of Germs: The Persistent Association of Immigrants and Disease in American Society". In: *Milbank Q* 80, pp: 757–788.

Marmot, Michael. (2017): "Social justice, epidemiology and health inequalities". In: *Eur J Epidemiol* 32. N. 7, pp. 537–546.

Martinez, Omar; Wu, Elwin; Sandfort, Theo; et al. (2015): "Evaluating the Impact of Immigration Policies on Health Status Among Undocumented Immigrants: A Systematic Review". In: *J Immigr Minor Health* 17. N. 3, pp. 947–970.

Masera, Luca. (2018): "L'incriminazione dei soccorsi in mare: dobbiamo rassegnarci al disumano?". In: *Questione Giustizia* 2. Pp. 225–238.

Mauss, Marcel. (1966): *The gift; forms and functions of exchange in archaic societies.* London: Cohen & West.

McAuley, Andrew. (2014): "Digital health interventions: widening access or widening inequalities?". In: *Public Health* 128. N. 12, pp. 1118–1120. doi: 10.1016/j.puhe.2014.10.008.

McGuire, Sharon; Martin, Kate. (2007): "Fractured migrant families: paradoxes of hope and devastation". In: *Family and Community Health* 30. Pp. 178–188.

MEDITERRANEA. (2022): *Report 2022.* https://mediterranearescue.org/wp-content/uploads/2022/02/MedReport_2022_01_ITA.pdf, visited on 13 July 2022.

MEDU [Medici per i Diritti Umani]. (2017): *Una malattia chiamata tortura. Rotte migratorie dall'africa sub-sahariana all'Europa.* http://www.mediciperidirittiumani.org/una-malattia-chiamata-tortura-rotte-migratorie-dallafrica-sub-sahariana-alleuropa/, visited on 13 July 2022.

MEDU [Medici per i Diritti Umani]. (2013) Arcipelago CIE. Indagine sui centri di identificazione ed espulsione italiani, available at: http://www.mediciperidirittiumani. org/pdf/ARCIPELAGOCIEsintesi.pdf, visited on 13 July 2022.

MELTINGPOT. (2021): *Diritti in rotta. Le navi quarantena come dispositivo di privazione della libertà personale.* https://www.meltingpot.org/Diritti-in-rotta-L-esperimento-delle-navi-quarantena-le.html#.YNIpPzJR0OQ, visited on 13 July 2022.

Mengesha, Zelalem; Alloun, Esther; Weber, Danielle; Smith, Michell; Harris, Patrick. (2022): "'Lived the Pandemic Twice': A Scoping Review of the Unequal Impact of the COVID-19 Pandemic on Asylum Seekers and Undocumented Migrants". In: *Int J Environ Res Public Health* 19. N. 11, 6624, doi: 10.3390/ijerph19116624.

Mocellin Raymundo, Marcia. (2011): "Uma aproximação entre bioética e interculturalidade em saúde a partir da diversidade". In: *Revista HCPA* 31. N. 4, pp. 494–499.

Moor, Irene; Spallek, Jacob; Richter, Matthias. (2017): "Explaining socioeconomic inequalities in self-rated health: a systematic review of the relative contribution of material, psychosocial and behavioural factors". In: *J Epidemiol Community Health* 71. N. 6, pp. 565–575.

MSF [Médecins Sans Frontières]. (2016): *Traumi Ignorati. Richiedenti asilo in Italia: un'indagine sul disagio mentale e l'accesso ai servizi sanitari territoriali.* http://archivio.medicisenzafrontiere.it/pdf/Rapp_Traumi_Ignorati_140716B.pdf, visited on 13 July 2022.

Mukumbang, Ferdinand C. (2021): "Pervasive systemic drivers underpin COVID-19 vulnerabilities in migrants". In: *Int J Equity Health* 20. N. 1, 146, doi: 10.1186/s12939-021-01487-2.

OHCHR [Office of the High Commissioner for Human Rights, United Nations]. (2020a): *COVID-19: States Should Not Abuse Emergency Measures to Suppress Human Rights.* https://www.ohchr.org/EN/NewsEvents/Pages/DisplayNews.aspx?NewsID=25722, visited on 13 July 2022.

OHCHR [Office of the High Commissioner for Human Rights, United Nations]. (2020b): *Statement on derogations from the Covenant in connection with the COVID-19 pandemic.* https://www.ohchr.org/Documents/HRBodies/CCPR/COVIDstatementEN.pdf, visited on 13 July 2022.

Oppel, Richard A. Jr.; Gebeloff, Robert; Lai, Rebecca K. K.; Wright, Will;. Smith, Mitch. (2020): "The Fullest Look Yet at the Racial Inequity of Coronavirus". In: *New York Times*, 05 July 2020. https://www.nytimes.com/interactive/2020/07/05/us/coronavirus-latinos-african-americans-cdc-data.html, visited on 13 July 2022.

Paccoud, Ivana; Baumann, Michelle; Le Bihan, Etienne; et al. (2021): "Socioeconomic and behavioural factors associated with access to and use of Personal Health Records". In: *BMC Med Inform Decis Mak* 21. N. 1, 18, doi: 10.1186/s12911-020-01383-9.

Parkin, Joanna. (2013): "The Criminalisation of Migration in Europe. A State-of-the-Art of the Academic Literature and Research". In: *CEPS Paper in Liberty and Security in Europe* 61. https://www.ceps.eu/ceps-publications/criminalisation-migration-europe-state-art-academic-literature-and-research/, visited on 12 July 2022.

Pfortmueller, Carmen A.; Schwetlick, Miriam; Mueller, Thomas; et al. (2016): "Adult Asylum Seekers from the Middle East Including Syria in Central Europe: What Are Their Health Care Problems?". In: *PLoS One* 11. N. 2, e0148196.

RED ACOGE. (2015): *Los efectos de la exclusión sanitaria en las personas inmigrantes más vulnerables.* http://redacoge.org/mm/file/2015/Jurídico/Informe%20Sanidad %20RED_ACOGE.pdf, visited on 13 July 2022.

REDER. (2017): *La salud en los márgenes del sistema.* https://www.reder162012.org, visited on 13 July 2022.

Reyneri, Emilio. (2011): *Sociologia del mercato del lavoro. Le forme dell'occupazione.* Bologna: Il Mulino.

Rigo, Enrica. (2007): *Europa di confine. Trasformazioni della cittadinanza nell'Unione allargata*. Roma: Meltemi.

Robinson, Laura; Schulz, Jeremy; Khilnani, Aneka; et al. (2020): "Digital inequalities in time of pandemic: COVID-19 exposure risk profiles and new forms of vulnerability". In: *First Monday* 25. N. 7, ff10.5210/fm.v25i7.10845 ff.

Sanfelici, Mara. (2021): "The Impact of the COVID-19 Crisis on Marginal Migrant Populations in Italy". In: *American Behavioral Scientist* 65. N. 10, pp. 1323–1341.

Santoro, Emilio. (2006): "Dalla cittadinanza inclusiva alla cittadinanza escludente: il ruolo del carcere nel governo delle migrazioni". In: *Diritto e questioni pubbliche* 6. Pp. 39–79.

Santoro, Emilio. (2009). "La regolamentazione dell'immigrazione come questione sociale". In Santoro, Emilio (ed), *Diritto come questione sociale*. Giappichelli, Turin.

Santoro, Emilio. (2021): "Danzando sui gusci d'uovo. Indicazioni per rendere più inclusive le procedure previste dall'art. 103 DL 34/2010". In: *L'Altro Diritto ODV*. http://www.adir. unifi.it/odv/adirmigranti/danzando-gusci-uovo.pdf, visited on 13 July 2022.

Sayad, Abdelmalek. (1999): *La Double Absence. Des illusions de l'émigré aux souffrances de l'immigré*. Seuil: Paris.

Sayad, Abdelmalek. (2006): *L'immigration ou les paradoxes de l'altérité. L'illusion du provisoire*. Paris: Raisons d'Agir.

Soto, Sara M. (2009): "Human migration and infectious diseases". In: *Clinical Microbiology and Infection* 15. Suppl. 1, pp: 26–28.

Thomson, Stephen; Ip, Eric C. (2020): "COVID-19 emergency measures and the impending authoritarian pandemic". In: *Journal of Law and the Biosciences* 7. N. 1, pp: 1–33.

Tognetti Bordogna, Mara. (2013): "Nuove diseguaglianze in salute: il caso degli immigrati". In: *Cambio* 3. N. 5, pp: 59–72.

Torre, Valeria. (2019): "Lo sfruttamento del lavoro. La tipicità dell'art. 603-bis cp tra diritto sostanziale e prassi giurisprudenziale". In: *Questione Giustizia* 4. Pp. 90–97.

Turner, Leigh. (2004): "Bioethics needs to rethink its agenda". In: *BMJ* 328. N. 7432, 175. doi: 10.1136/bmj.328.7432.175.

Turner, Leigh. (2005): "Bioethics, Social Class, and the Sociological Imagination". In: *Cambridge Quarterly of Healthcare Ethics* 14. Pp. 374–378.

UN [United Nations]. (2000a): *Protocol against the Smuggling of Migrants by Land, Sea and Air, supplementing the United Nations Convention against Transnational Organized Crime*. https://www.unodc.org/documents/southeastasiaandpacific/2011/04/somind onesia/convention_smug_eng.pdf, visited on 13 July 2022.

UN [United Nations]. (2000b): *Protocol to Prevent, Suppress and Punish Trafficking in Persons, especially Women and Children, supplementing the United Nations Convention against Transnational Organized Crime*. https://www.osce.org/odihr/19223?download= true, visited on 13 July 2022.

UN [United Nations]. (2010): *UNAIDS Outcome Framework 2009–2011*. http://www.unaids.org/ sites/default/files/sub_landing/files/20100728_HR_Poster_en.pdf, visited on 13 July 2022.

UN [United Nations]. (1951): *Convention and protocol relating to the status of refugees*. https://www.unhcr.org/1951-refugee-convention.html, visited on 13 July 2022.

UNESCO. (2001): "Declaration on Cultural Diversity". In: *Cultural Diversity Series* 1. http://un esdoc.unesco.org/images/0012/001271/127162e.pdf, visited on 13 July 2022.

Valeriani, Giuseppe; Sarajlic Vukovic, Iris; Lindegaard, Tomas; et al. (2020): "Addressing Healthcare Gaps in Sweden during the COVID-19 Outbreak: On Community Outreach and Empowering Ethnic Minority Groups in a Digitalized Context". In: *Healthcare* 8. N. 4, 445. doi: 10.3390/healthcare8040445.

van Deursen, Alexander J. A. M. (2020): "Digital Inequality During a Pandemic: Quantitative Study of Differences in COVID-19-Related Internet Uses and Outcomes Among the General Population". In: *JMIR* 22. N. 8, e20073. doi: 10.2196/20073.

WHO. (1998): *A Conceptual Framework for Action on the Social Determinants of Health*. http://www.who.int/social_determinants/resources/csdh_framework_action_ 05_07.pdf, visited on 13 July 2022.

WHO. (2003): *Social determinants of health The Solid Facts*. http://www.euro.who.int/__data/assets/pdf_file/0005/98438 /e81384.pdf, visited on 13 July 2022.

WHO. (2013): *Migración internacional, salud y derechos humanos*. http://www.ohchr.org/Documents/Issues/Migration/WHO_IOM_UNOHCHRPublication_sp.pdf, visited on 13 July 2022.

WHO. (2017a): *Migrations and health: key issues*. http://www.euro.who.int/en/health-topics/health-determinants/migration-and-health/migrant-health-in-the-european-region/migration-and-health-key-issues, visited on 13 July 2022.

WHO. (2017b): "A review of evidence on equitable delivery, access and utilization of immunization services for migrants and refugees in the WHO European Region". In: *Health Evidence Network Synthesis Report* 53. http://www.euro.who.int/en/publications/abstracts/review-of-evidence-on-equitable-delivery-access-and-utilization-of-immuniza tion-services-for-migrants-and-refugees-in-the-who-european-region-a-2017, visited on 13 July 2022.

Worthing, Kitty; Galaso, Marta M.; Wright, Johanna K.; Potter, Jessica. (2021): "Patients or passports? The 'hostile environment' in the NHS". In: *Future Healthc J* 8. N. 1, pp. 28–30. doi: 10.7861/fhj.2021–0007.

Zorzella, Nazzarena. (2020): "Regolarizzazione 2020, una prevedibile occasione perduta. Alcune delle principali criticità". In. *Critica del diritto*. https://rivistacriticadeldiritto.it/?p=1458, visited on 13 July 2022.

Authors Biographical Information

Bhuiyan Md Sohrab Hossain has a Master in English Literature and Language from National University of Bangladesh, a Master on Islamic Studies from National University of Bangladesh, a Bachelor in Education from Government Teachers Training College. He worked as an English lecturer and coordinator of English courses in schools and colleges in Dhaka. In 2012, he migrated to Italy where he currently lives.

Carlo Botrugno, PhD, BSW, LLM, LLB, is Assistant Professor at the Department of Legal Sciences at University of Florence; founder and coordinator of the Research Unit on Everyday Bioethics and Ethics of Science (RUEBES); responsible to the Law Clinic in Bioethics at the School of Law of University of Florence; editor of L'Altro Diritto Journal (ISSN: 1827–0565). He is a consultant at the European Oncological Institute (IEO) in Milan; a member of the Clinical Ethics Committee (CEC) at AUSL-IRCCS Reggio Emilia Research Hospital; a lecturer for the Master's course on eHealth and Telemedicine Management at Rome Business School (RBS); and a guest lecturer in PhD, Master's and graduate programs on bioethics, biolaw, digital health and telemedicine in several countries in Europe and Latin America. He was a consultant at the Bioethics Unit at AUSL-IRCCS Reggio Emilia Research Hospital (2021); a visiting professor at Pernambuco Federal University (2019); a visiting professor at University of Tilburg (2019); a postdoctoral researcher at the Institute of Health Care Ethics at Slovak Medical University (2018); a visiting scholar at the Center for Social Studies (CES) at University of Coimbra (2015); a visiting scholar at the Experimental Research Unit on Bioethics and Ethics of Science (LAPEBEC) at Clinical Hospital in Porto Alegre (2015); a visiting scholar at Uniritter dos Reis in Porto Alegre (2015); and a researcher and teaching fellow at Interdepartmental Research Centre on History of Law, Philosophy and Sociology of Law and Legal Informatics (currently Alma-AI) at University of Bologna (2009–2015).

Celia Mariana Barbosa de Souza is a nurse graduate of University of Vale do Rio dos Sinos (UNISINOS) (1982) and has a specialization in Hospital Administration from the Pontifical Catholic University of Rio Grande do Sul (PUCRS) (1986); a specialization in Nephrology from the Federal University of Rio Grande do Sul (UFRGS) (2005); a Master's in Nephrology from the PUCRS (2012) with a focus on chronic kidney disease and the black population. Since 2018, she has been enrolled in the Medical Sciences Programme at UFRGS at the PhD level with a project that aims to develop a more comprehensive understanding of the correlations among chronic kidney disease, APOL1 risk variants, and the black population. Since 1986, she has been a nurse at Clinical Hospital in Porto Alegre (HCPA) and is currently working in the health unit. She is a member of the HCPA committee for the evaluation of candidates willing to apply for job quotas reserved to black people. She is also a member of the social collective MUSAS (*Mulheres na Universidade e na Saúde*) at UFRGS.

Cláudio Lorenzo has a Master's in Health and Medicine with a specialization in Bioethics from the Federal University of Bahia; he has been awarded a PhD in Ethics of Applied Clinical Sciences from the University of Sherbrooke, Canada. He is currently an Associate Professor at the Department of Public Health at University of Brasília (UnB) and Professor of the Post-Graduate Programmes in Bioethics and Public Health at the same university. He is the coordi-

https://doi.org/10.1515/9783110765120-014

nator of the Center for Studies in Public Health of the Federal District (Brazil) and is currently enrolling in a postdoctoral internship at the Postgraduate Program in Sociology at UnB. He has been the president of the Brazilian Society of Bioethics (from 2011 to 2013) and a member of the Advisory Committee of Latin American and Caribbean Network on Bioethics (from 2010 up to 2022). His experience is related to the areas of Bioethics and Public Health, and he mostly teaches, researches and consults on the following topics: research ethics, bioethics, and applied social sciences in the health field.

Fernanda Sales Luiz Vianna is Assistant Professor with the Genetics Department, Biosciences Institute of the Federal University of Rio Grande do Sul (UFRGS), Porto Alegre; Head of the Laboratory of Genomics Medicine, Experimental Research Service at Clinical Hospital of Porto Alegre (HCPA). She is a graduate in Biological Sciences of Pontifical Catholic University of Rio Grande do Sul, Porto Alegre (2005), and she has a Master's in Genetics and Molecular Biology at UFRGS (2008) and a PhD in Sciences, Genetics, and Molecular Biology at UFRGS (2013). She carried out a postdoc in Epidemiology at UFRGS (2016). From 2016 to 2019, she was ai research assistant with the Bioethical Division of HCPA. Currently, she is Assistant Professor at UFRGS and head of the Research Group on Immunogenetics Research; she works on human genetics, immunogenetics, and medical and population genetics.

Francisco Veríssimo Veronese is Associate Professor with the Internal Medicine Department, Faculty of Medicine at Federal University of Rio Grande do Sul (UFRGS); Assistant Physician at the Nephrology Department of Hospital de Clínicas de Porto Alegre (HCPA). He graduated in Medicine from UFRGS (1982), completed a Master's in Nephrology at UFRGS (1990) working with renal lithiasis, and has a PhD in Medical Sciences from UFRGS (2001) with a dissertation on "Prevalence and Immunohistochemical Findings of Subclinical Kidney Allograft Rejection and its Association with Graft Outcome". He carried out a postdoc at Harvard University, Massachusetts General Hospital (MA), Boston, working as a Fellow in Renal Pathology and Immunopathology (2005–2006). From 1994 to 2007, he worked as a contracted physician in the Nephrology Department, HCPA, focusing on the following renal diseases: renal transplantation, acute and chronic kidney diseases, hemodialysis and peritoneal dialysis. Since 2008, he has been Associate Professor at UFRGS, working on clinical nephrology, glomerulopathies, and renal immunopathology. He is a coordinator of the Research Group on Glomerulopathies by the Research Incentive Fund (FIPE) of HCPA and the Brazilian Research Council (CNPq).

Inês Faria is a researcher at the CSG-SOCIUS/ISEG, University of Lisbon. She has a PhD in Medical Anthropology from the University of Amsterdam, focusing on reproductive health, reproductive technologies and therapeutic navigations of Mozambican women and couples in their quest to overcome infertility in Mozambique and South Africa. Since 2016, she has been conducting research concerning money and entrepreneurship in the context of crypto currencies and blockchain technology. Her latest work focuses on financial uses of blockchain technology, women, and financial inclusion development programs. She was involved in the project Finance Beyond Fact and Fiction, which explored financial changes and continuities in Europe after the 2008 crisis and is now developing the project Informal Businesses and Informed Financial Creativity to investigate digital microfinance and the financial ecologies of entrepreneurial women in Maputo. Meanwhile Inês continues to focus on the relations between technology and society in the areas of money, finance, livelihoods and healthcare in the European and sub-Saharan African contexts.

Josimário João da Silva has a Master's and a PhD in Surgery from Catholic Pontifical University of Rio Grande do Sul (PUCRS); a specialization in Clinical Bioethics at Fundación Ciencia de la Salud; a Bachelor's in Law from the Immaculate Catholic Faculty of Recife; and a Master's in Medical Law from the University of Santo Amaro. He carried out a postdoc in Bioethics at São Camilo University. Currently, he is Professor at the Center for Medical Sciences of the Federal University of Pernambuco; Head of the Clinical Bioethics Service of the Hospital das Clínicas of the Federal University of Pernambuco; National Coordinator of the Bioethics Network Brazil; President of the Pernambuco Institute of Bioethics and Biolaw; President of the Brazilian Academy of Clinical Bioethics; and a Master's student in Bioethics at Fundación Ciencia de la Salud.

Judith van de Kamp is a medical anthropologist and assistant professor global health at the Julius Center for Health Sciences and Primary Care, University Medical Center Utrecht. She received her PhD from the University of Amsterdam. Her research focuses on equity in partnerships and relationships for health, and in particular on the involvement of international health workers and health students who temporarily work in health settings in low- and middle-income countries (LMICs). Her work pertains to the White Savior Industrial Complex, power imbalances, colonial mentality, feelings of superiority, responsible behavior, (im)personal authority and hierarchy, and intercultural competences. Beside research, Judith is involved in teaching and developing interdisciplinary global health education for various groups of students. She is passionate about sharing her research results with a variety of audiences, ranging from medical students to NGO workers, with the main goal of encouraging people to reflect critically on their role in LMICs. She wrote a book about that topic for a readership of the Dutch general public (De derde wereld op je cv, 2019).

Karla Costa is a nutritionist at the Federal University of Pernambuco (2008–2011) and a specialist in Family Health from the University of Pernambuco (2012–2014). She has a Master's in Child and Adolescent Health from the Federal University of Pernambuco (2014–2016); has been a researcher with a doctoral internship at the Center for Social Studies (CES) of the University of Coimbra (2019–2021); has a PhD in Public Health (2017–2022); she is a member of the Association SaMaNe (Saúde das Mães Negras) (2020-current), which aims to understand the experiences of black and Afro-descendant women in Portugal concerning sexual and reproductive health care and to broaden the political, academic and social debate pertaining to the theme. Her research interests are related to post-colonialism, primary health care, nutrition and collective health, reproductive health and justice, structural violence, racism, and intersectionality.

Katya Gibel Mevorach is Professor of Anthropology and American Studies at Grinnell College. She earned her PhD in Cultural Anthropology from Duke University and received a BA and MA in African Studies from Hebrew University of Jerusalem in Israel. Under the name, Katya Gibel Azoulay, she is author of "Black, Jewish and Interracial: It's Not the Color of Your Skin but the Race of Your Kin, and Other Myths of Identity" (Duke University Press, 1997), an explication of identities in an historical context. Her current research and teaching focus on identities and social science categories, "race, racism and racecraft", and their articulation in academic curricula as well as interrogating the commodification of diversity in higher education. Journals in which her articles, review essays and position papers have appeared include American Anthropologist, American Ethnologist, Biography, Developing World Bioethics, Identities, Cultural Studies, Research in African Literatures, Noga (Israeli Feminist Journal) and The Jerusalem Post (Israel). Gibel Mevorach has dual American–Israeli citizenship. She moved to Israel in 1970 after graduating from The Brearley School in New York, returned to the US in 1991 to pursue her doctoral studies, and was invited to join Grinnell College as a Scholar-in-Residence in January 1996. Gibel Mevorach has served as Chair of the Africana Studies Concentration from 1996–2000, and in 2003, she helped initiate the transition of Africana Studies into an expanded American Studies Concentration, which she chaired between 2004 and 2005 and 2015–19.

Laura Brito, has a Master's in medical anthropology and is a PhD candidate in post-colonialism and global citizenship at the Faculty of Economics/Center for Social Studies (FEUC/CES) of the University of Coimbra. She is a member of the social collective SaMaNe (Saúde das Mães Negras) (Health of the Black Mothers), which aims to understand the experiences of black and Afro-descendant women in Portugal concerning sexual and reproductive health care and to raise awareness of the importance of culturally competent health care. Her PhD project aims to understand how race, class and gender oppression interfere in the pregnancy, childbirth, and postpartum experiences of black and Afro-descendant women in Lisbon. Her research interests focus on postcolonialism, reproductive health, reproductive justice, racism, human rights and intersectionality.

Lucia Re is Associate Professor of Philosophy of Law at the University of Florence, Italy. In 2021, she obtained the national scientific qualification as full Professor. She is part of the Teaching Board of the PhD in Legal Sciences of the University of Florence. She is the director of "Jura Gentium. Journal of Philosophy of International Law and Global Politics". She is part of the directive committee of the Journal "Studi sulla questione criminale". She has been deputy director of the Journal "L'altro diritto" (2017–2022) and she is a current member of its Editorial Board. She is member of the editorial board of the journals, "Revista Direito Mackenzie" and "Revista de Direitos e Garantias Fundamentais". She is a member of the Board of Directors of Center Jura Gentium and of the Scientific Council of L'altro diritto Inter-university research center (ADir), and the Research Unit on Everyday Bioethics and Ethics of Science. She is also co-coordinator of the Inter-University Working Group on Women's Political Subjectivity. She has been part of several research teams at University of Florence. She was responsible for the research unit of the University of Florence in the PRIN 2015 on the "Subject of right and vulnerability (2017–2020)". Her research focuses on fundamental rights, discrimination, and gender.

Marcia Mocellin Raymundo is a biologist at the Bioethics Department of Clinical Hospital of Porto Alegre, Brazil. She has a Master's in Physiological Sciences (2000) at Federal University of Rio Grande do Sul (UFRGS), Brazil and a PhD in Sciences of Gastroenterology (2007) at UFRGS. She carried out a postdoc (2010–2011) in Bioethics at the National Autonomous University of Mexico (UNAM). Currently, she is a member of the Advisory Committee of the Latin America and the Caribbean Bioethics Network, created by UNESCO; a lecturer for the course Bioethics, Diversity and Gender of the post-graduate program in Gynecology and Obstetrics at UFRGS. Her research interests focus on social bioethics, diversity, racism, interculturality and laicity.

Nchangwi Syntia Munung is a researcher at the Division of Human Genetics, University of Cape Town, South Africa. She has a PhD in Medicine from the University of Cape Town, focusing on how theories of global health justice and the African ethic of Ubuntu can inform equitable global governance of genomics. Nchangwi is currently working on issues related to the governance of global health research programs, the ethics of infectious disease control and management, public engagement for genomics and data science in health-related, ethical, and social issues related to emerging biotechnologies.

Robin L. Pierce is Professor of AI and the Law at the University of Exeter. She earned her PhD in Health Policy from Harvard University and received a JD (Juris Doctor) from the University of California, Berkeley and a BA from Princeton University. Her current research focuses on legal, policy-related, and ethical issues pertaining to the translation of AI-driven technologies in healthcare and medical science, including the life sciences in addition to her ongoing research agenda concerning "Interrupting Pathways to Health Inequities", represented by her chapter in this volume, "Interrupting Technological Pathways to Health Inequities". Ranging from understanding the dynamics and implications of the early detection of disease to the active engagement of norms in the context of public and clinical health, her work seeks to identify sustainable ways of achieving optimal and equitable health outcomes while preserving and promoting conditions for human flourishing and self-actualization. Her publications have appeared in such journals as Health Promotion, The Lancet Neurology, Biotechnology, European Data Protection Law, and Social Science and Medicine. She has been an invited speaker on AI and Robotics in Healthcare at the European Parliament and has served on advisory boards for biotech and biomedical research projects. She has been affiliated with universities in both Europe and North America and currently resides in the UK.

Ronaldo Piber is a lawyer specializing in medical and health law. He has a Master's in Bioethics from Universidad Europea del Atlántico and a Master's in Medical Law from Santo Amaro University. He is a Member of the International Forum of Teachers of the UNESCO Chair of Bioethics; an Associate Member of the World Association for Medical Law (WAML); and the President of the Bioethics and Biolaw Commission of the Brazilian Bar Association, Pinheiros Subsection – São Paulo/SP (triennium 2022/2024).

Index

https://doi.org/10.1515/9783110765120-015

www.ingramcontent.com/pod-product-compliance
Lightning Source LLC
Chambersburg PA
CBHW072300210326
41519CB00057B/2362